T0248336

International Space Traffic Management

Charting a Course for Long-Term Sustainability

BRUCE MCCLINTOCK, DOUGLAS C. LIGOR, DAN MCCORMICK, MARISSA
HERRON, KOTRYNA JUKNEVICIUTE, THOMAS VAN BIBBER, KATIE FEISTEL,
AKHIL RAO, ADI RAO, TAYLOR GROSSO, MICHAEL FENNER, HANJUN LEE,
ABDULLAH AR RAFEE, TOMÁS URBINA

Prepared for the Office of the Secretary of Defense
Approved for public release; distribution is unlimited

NATIONAL DEFENSE RESEARCH INSTITUTE

For more information on this publication, visit **www.rand.org/t/RRA1949-1**.

About RAND

The RAND Corporation is a research organization that develops solutions to public policy challenges to help make communities throughout the world safer and more secure, healthier and more prosperous. RAND is nonprofit, nonpartisan, and committed to the public interest. To learn more about RAND, visit www.rand.org.

Research Integrity

Our mission to help improve policy and decisionmaking through research and analysis is enabled through our core values of quality and objectivity and our unwavering commitment to the highest level of integrity and ethical behavior. To help ensure our research and analysis are rigorous, objective, and nonpartisan, we subject our research publications to a robust and exacting quality-assurance process; avoid both the appearance and reality of financial and other conflicts of interest through staff training, project screening, and a policy of mandatory disclosure; and pursue transparency in our research engagements through our commitment to the open publication of our research findings and recommendations, disclosure of the source of funding of published research, and policies to ensure intellectual independence. For more information, visit www.rand.org/about/research-integrity.

RAND's publications do not necessarily reflect the opinions of its research clients and sponsors.

Published by the RAND Corporation, Santa Monica, Calif.
© 2023 RAND Corporation
RAND® is a registered trademark.

Library of Congress Cataloging-in-Publication Data is available for this publication.
ISBN: 978-1-9774-1141-9

Cover design by Rick Penn-Kraus

Cover photos: Space: dottedhippo/Getty Images/iStockphoto; left satellite: PhonlamaiPhoto/Getty Images/ iStockphoto; right satellite: alxpin/Getty Images/iStockphoto; Death Star: Andrew Forrester/The Noun Project.

About This Report

As outer space becomes more congested, contested, and competitive, the risks to space safety, security, and sustainability heighten. Against this backdrop, we used a review of relevant literature and official documents and interviews and workshops with subject-matter experts to identify possible lessons for future space traffic management from past approaches to international traffic management or common resource management.

This research should be of interest to national and international government and military leaders and policymakers, space industry leaders and organizations, public and private interest groups and institutions involved in space, those working in academia, and all those concerned with developing formal government inputs to the issue of space traffic management. The examination of traffic coordination in other domains, as well as the recommendations resulting from an analysis of these domains, will be most useful to established and aspirant spacefaring powers and policymakers, as well as space industry participants and operators. The research reported here was completed in January 2023 and underwent security review with the sponsor and the Defense Office of Prepublication and Security Review before public release.

RAND National Security Research Division

This research was conducted within the International Security and Defense Policy Program of the RAND National Security Research Division (NSRD), which operates the RAND National Defense Research Institute (NDRI), a federally funded research and development center (FFRDC) sponsored by the Office of the Secretary of Defense, the Joint Staff, the Unified Combatant Commands, the Navy, the Marine Corps, the defense agencies, and the defense intelligence enterprise. This research was made possible by NDRI exploratory research funding that was provided through the FFRDC contract and approved by NDRI's primary sponsor.

For more information on the RAND International Security and Defense Policy Center, see www.rand.org/nsrd/isdp or contact the director (contact information is provided on the webpage).

Acknowledgments

We are grateful to Jack Riley, Barry Pavel, and Mike Spirtas for their vision and efforts to better connect the RAND Corporation's extensive space-related research through the RAND Space Enterprise Initiative. This report is a better product because of the thoughtful and detailed reviews provided by Ruth Stilwell and James Black. The report benefited from inputs from numerous workshop participants from several countries and various space organizations, multiple interview subjects, and informal consultants. Additionally, we thank Robert Murrett of Syracuse University and members of the faculty of the Department of Political Science at the United States Air Force for conducting forums in which concepts and ideas for this report were further developed. We alone are responsible for the final content of this report and any errors that may remain.

Summary

Space traffic management (STM) is one of the most pressing and challenging issues for the space domain. Space is a domain that all of humanity heavily relies on for critical services, benefits, and activities, which include defense and national security activities; position, navigation, and timing (PNT); satellite communications (SATCOM); internet service; television and cable broadcasting; international financial transactions; remote Earth sensing; weather monitoring and prediction; and scientific exploration and experimentation. These services and benefits are currently at significant risk because safely maneuvering satellites and other space objects to avoid colliding with other active objects and debris is becoming more difficult for space operators.

STM is currently an ad hoc process. This informal and ill-coordinated approach is likely to deteriorate in effectiveness as congestion in space increases.[1] As of December 2022, there were approximately 6,900 active satellites, more than 36,500 pieces of trackable debris (10 cm or larger), and approximately 1 million objects sized 1 cm to 10 cm in space—all of which could cause significant or catastrophic loss of valuable orbital assets in the event of cascading collisions (often referred to as the Kessler Syndrome).[2] Moreover, the number and value of space objects and the services derived from those objects are increasing rapidly. By some estimates, tens of thousands of additional satellites might occupy low earth orbit (LEO) by 2030, compounding the risks and threats to successful STM.[3] Additionally, having to maneuver to safety in a reactive manner because of a lack of effective STM imposes additional costs in terms of fuel, which necessarily shortens the lifespan of the satellites. These shortened lifespans not only reduce the return on investment of any such assets; they also risk creating more dangerous debris if those defunct satellites are not disposed of safely and sustainably.[4]

---

[1] Michael P. Gleason and Travis Cottom, *U.S. Space Traffic Management: Best Practices, Guidelines, Standards, and International Considerations*, Aerospace Corporation, August 2018, pp. 7–8.

[2] European Space Agency, "Space Debris by the Numbers," webpage, undated. Additionally, nontrackable debris measuring 1 mm to 1 cm, which could also damage or disable a space object, numbers approximately 1,300,000 pieces (Michael Dominguez, Martin Faga, Jane Fountain, Patrick Kennedy, and Sean O'Keefe, *Space Traffic Management*, National Academy of Public Administration, August 2020; European Space Policy Institute, *Space Environment Capacity: Policy, Regulatory and Diplomatic Perspectives on Threshold-Based Models for Space Safety and Sustainability*, April 11, 2022).

[3] Inmarsat, *Space Sustainability Report: Making the Case for ESG Regulation, International Standards and Safe Practices in Earth Orbit*, June 22, 2022, pp. 16–20.

[4] See Jason Rainbow, "Connecting the Dots: Improving Satellite Collision Predictions for Efficient Space," *SpaceNews*, June 23, 2022 ("Depending on these factors for a given operator, they could perform anywhere from only a few [collision avoidance] maneuvers per year, all the way to multiple maneuvers per week, per satellite,"); Tereza Pultarova, "SpaceX Starlink Satellites Responsible for over Half of Close Encounters in Orbit, Scientist Says," *Space.com*, August 20, 2021 ("an operator managing about 50 satellites will receive up to 300 official conjunction alerts a week").

At its core, the STM problem is more of a governance problem than a technical one.[5] To avoid interference and collision, space actors must coordinate in the form of communications; data and information exchange; situational awareness; conflict resolution; and defined processes and procedures to determine and adjudicate who will move, when the move will occur, and how the move will be executed to ensure safety, all of which are governance functions. In this report, we aim to offer analysis, insights, and recommendations for the development of a governance framework for international STM.

To accomplish this, we examine both the maritime and air domains in depth. Both offer successful and longstanding governance mechanisms from which the space domain might draw lessons, particularly lessons related to traffic management. Additionally, we explore other domains that do not have traffic management as a discrete problem set but involve governance mechanisms which could be analogous to space (e.g., telecommunications, the internet, and international banking). Although these domains are not physically analogous to space, we find that many of their governance processes, procedures, mechanisms, and institutions are generally applicable to addressing such challenges as STM.[6]

To inform this analysis, we conducted a thematic literature review to identify and collect publicly available materials, documents, data, and information relevant to the topic of international governance organizations and, specifically, issues related to traffic management.[7] Additionally, we conducted workshops with recognized international scholars, subject-matter experts, and government officials. To ensure a diversity of views, our workshops were

---

[5] Generally, *governance* is defined as the ability of ability of an authoritative entity to make and enforce rules and to deliver services. See Francis Fukuyama, "What Is Governance?" *Governance*, Vol. 26, No. 3, July 2013, p. 350. We include in this definition the ability of an authoritative entity to "direct and control the actions or conduct of" entities in the system "either by established laws [rules]" and to "direct and control, rule, or regulate" by agreed upon rules, precedents, and principles. See generally the definition of "govern," *Black's Law Dictionary*, 6th ed., 1990, p. 695.

[6] We note that many entities and subject-matter experts have also examined these and other domains for applicability to space, and we attempted to review and summarize those previous efforts as part of our research. Many of those works focus on specific technologies, policies, or processes. Our approach differs in that we sought to focus on governing institutions and structures and the characteristics of those structures as they related to governance (e.g., organization, functions, roles, responsibilities, bureaucratic mechanisms and procedures).

[7] We conducted open online searches using various search engines (Google, Google Scholar, Research-Gate, HeinOnline, LexisNexis). We also conducted searches of websites associated with international air, maritime, telecommunications, internet, and banking governance (e.g., UN.org, ICAO.org, IMO.org, ITU. int, ICAAN.org, SWIFT.com). To select source material with the most relevance, materiality, and importance, we applied keyword terms and phrases related to governance and traffic management to our searches and reviewed abstracts, summaries, and full documents to determine the frequency and applicability of the keyword terms. Keyword search terms and phrases consisted of the following: international governance, international governmental organizations (IGOs), traffic, traffic management, air traffic, air traffic management, air traffic control, maritime traffic, maritime traffic management, governance, governance processes, governance procedures, governance mechanisms, governance organizations, treaty mechanisms, Chicago Convention, ICAO, ICAN, IMO, UNCLOS, ITU, ICANN, and SWIFT.

broken down into three sessions. Each session covered a geographic area and corresponding set of space stakeholders: Europe, Asia-Pacific, and the United States.

Key Insights

Our research generated six key insights:

- The international community is at a tipping point for STM. Our research strongly indicated that the space domain's safety and sustainability are under clear and present threat from debris and congestion. Because of this threat and because the space domain involves international actors (or entities under the jurisdiction and control of states), an international organization is needed to conduct STM. We refer to such an organization as an international space traffic management organization (ISTMO).
- There is sufficient research, which has been conducted by government, nonstate, and private entities, to justify the creation of an ISTMO.
- This ISTMO must have sufficient authority and jurisdiction and must ensure that the technical coordination and collaboration between states, industry, and other stakeholders necessary for successful STM governance and operations occur.
- Without sufficient authority and jurisdiction and implementation of coordination and collaboration, an ISTMO might lack the legitimacy to endure and conduct effective STM.
- An ISTMO must also be able to facilitate and incorporate bottom-up development of STM rules and activities that includes governments, industry, and nongovernmental organizations.
- To accomplish these objectives, an ISTMO will need to be funded, staffed, and resourced to maintain the necessary level of expertise and a dynamic operational tempo.

Recommendations

Our research generated four recommendations:

- The first step toward an ISTMO that can successfully universalize STM rules across all space stakeholders and provide STM operations that are predictable, reliable, sustainable, and safe is for the international community to come to an agreement that the creation of an ISTMO is the best course of action. To accomplish this, we recommend that spacefaring and nonspacefaring states call for an ISTMO convention to be held at the United Nations (UN).
- We recommend that this convention focuses on the current body of research that supports the creation of an ISTMO drawing from the analogous aspects and lessons learned from other domains, particularly the maritime and air domains.

- We also recommend that current space stakeholders ensure both the gathering of international experts on STM and the continued growth of institutional expertise on STM. This expertise is needed to ensure that both a convention and any nascent organization it might create are equipped to be legitimate, effective, and long-standing.
- Finally, we recommend that additional research be done to develop a source of sustainable and equitable funding for the ISTMO. We address traditional funding mechanisms common to the UN system and include a discussion of some nontraditional options. Although these nontraditional options, such as orbital-use fees or a tradeable bond system, are underdeveloped, they offer further areas of research and discussion for space domain stakeholders to consider.

Contents

Figures and Tables

Figures

Tables

CHAPTER 1

Introduction

Since approximately 2000, the boom in space accessibility and interest has fueled tremendous growth in the volume and variety of satellites orbiting Earth, as well as the data and services those satellites provide.[1] Space is relied on for critical services, benefits, and activities, including defense and national security activities; position, navigation, and timing (PNT); satellite communications (SATCOM); internet service; television and cable broadcasting; international financial transactions; remote Earth sensing; weather monitoring and prediction; and scientific exploration and experimentation. Although the ongoing New Space Era has brought significant advancements in security, connectivity, prosperity, and collective action, the advantages of this burgeoning space economy come with substantial and increasingly urgent risks and challenges.

The growing number of objects in orbit has significantly increased the potential for overcrowding, debris creation, and, ultimately, collisions as the most useful orbital altitudes steadily approach the limit of their carrying capacities.[2] As of December 2022, there were approximately 6,900 active satellites, more than 36,500 pieces of trackable debris (10 cm or larger), and approximately 1 million objects sized 1 cm to 10 cm in space.[3] Along with the

---

[1] For this report, we consider *space* to be above the modern Karman Line (100 km above mean sea level) when discussing space traffic and space traffic management (STM). However, existing STM scholarship concerning activity between 60,000 ft (Flight Level 600) and the Karman Line, as well as the recent U.S. response to a Chinese spy balloon, suggest that a centralized STM system will likely require a robust definition of space and that cross-domain lessons can be learned even from how to define key terms. See Theresa Hitchens, "Balloons vs. Satellites: Popping Some Misconceptions about Capability and Legality," *Breaking Defense*, February 7, 2023b; Stephen Hunter, "Safe Operations Above FL600," paper presented at the Space Traffic Management Conference 2015: The Evolving Landscape, Daytona Beach, Fla., November 12–13, 2015.

[2] Jonathan McDowell, *Space Activities in 2022*, Jonathan's Space Report, January 3, 2023; Glenn Peterson, Marlon Sorge, and William H. Ailor, "*Space Traffic Management in the Age of New Space*, Aerospace Corporation, April 2018; Sébastien Rouillon, "A Physico-Economic Model of Low Earth Orbit Management," *Environmental and Resource Economics*, Vol. 77, No. 4, December 2020.

[3] European Space Agency, "Space Debris by the Numbers," webpage, undated. Additionally, nontrackable debris measuring 1 mm to 1 cm, which could also damage or disable a space object, numbers approximately 1,300,000 pieces; Michael Dominguez, Martin Faga, Jane Fountain, Patrick Kennedy, and Sean O'Keefe, *Space Traffic Management*, National Academy of Public Administration, August 2020; European Space Policy Institute, *Space Environment Capacity: Policy, Regulatory and Diplomatic Perspectives on Threshold-Based Models for Space Safety and Sustainability*, April 11, 2022.

increasing quantity of space objects and activities, new types of near-Earth operations—mega-constellations, space tourism, in-orbit servicing and manufacturing, space tugs, active debris removal, just-in-time and AI-driven collision avoidance maneuvers, and more—have also upended the space domain, resulting in an orbital environment that is substantially more complex, congested, and risky in the New Space Era.[4]

Growing in tandem with this rise in orbital danger and complexity have been the calls from academics, members of civil society, policymakers, and industry leaders for improved governance of space traffic, particularly on an international level, to ensure the continued safety and sustainability of the final frontier. Such governance would amount to a global system of *STM*, most prominently defined by the International Academy of Astronautics (IAA) as a "set of technical and regulatory provisions for promoting safe access into outer space, operations in outer space, and return from outer space to Earth free from physical or radiofrequency interference."[5] Heightened awareness of STM needs and the urgency of worsening orbital conditions have escalated global space traffic debates from purely academic circles to the highest levels of government. STM is now an annual topic at United Nations Committee on the Peaceful Uses of Outer Space (UNCOPUOS) Legal Subcommittee proceedings.[6] However, little agreement exists on what structure international STM should eventually take or even what new steps, if any, should be taken in the short term in pursuit of reliable, global STM.

Currently, STM is an informal, ad hoc, and ill-coordinated process that is likely to prove ineffective as space becomes more congested.[7] It is estimated that tens of thousands of additional satellites will likely be launched into low earth orbit (LEO) by 2030, increasing the threat of collisions and risking the sustainable use of Earth's orbits.[8] Moreover, space operators increasingly have to maneuver their satellites to safely avoid collisions, imposing additional fuel costs and shortening the life of the satellites. These shortened lifespans not only

---

[4] International Academy of Astronautics, International Astronautical Federation, and International Institute of Space Law, *Cooperative Initiative to Develop Comprehensive Approaches and Proposals for Space Traffic Management (STM)*, September 17, 2022. We define the *New Space Era* as beginning from approximately the year 2000 to present. See Bruce McClintock, Katie Feistel, Douglas C. Ligor, and Kathryn O'Connor, *Responsible Space Behavior for the New Space Era: Preserving the Province of Humanity*, RAND Corporation, PE-A887-2, April 2021, p. 3.

[5] Corinne Contant-Jorgenson, Petr Lála, Kai-Uwe Schrogl, eds., *Cosmic Study on Space Traffic Management*, International Academy of Astronautics, 2006.

[6] United Nations Office for Outer Space Affairs, "Proposal for a Single Issue/Item for Discussion at the Fifty-Fifth Session of the Legal Subcommittee in 2016 on: 'Exchange of Views on the Concept of Space Traffic Management,'" April 2015.

[7] Michael P. Gleason and Travis Cottom, *U.S. Space Traffic Management: Best Practices, Guidelines, Standards, and International Considerations*, Aerospace Corporation, August 2018, pp. 7–8.

[8] Inmarsat, *Space Sustainability Report: Making the Case for ESG Regulation, International Standards and Safe Practices in Earth Orbit*, June 22, 2022, pp. 16–20.

increase costs, but also result in increasing debris because the defunct satellites cannot be disposed of sustainably.[9]

In this report, we posit that STM is primarily a governance challenge rather than a technical challenge.[10] To operate safely and sustainably in space, operators must coordinate and communicate; exchange data and information; enable situational awareness; avoid, mitigate, and resolve against conflict; and define processes and procedures to determine and adjudicate who will maneuver, when that maneuver will occur, and how the maneuver will be executed to ensure safety. All of these activities are also governance functions. We therefore seek to provide clarity to the ongoing space traffic solution debates by determining the most-advantageous features of a potential international STM system and the optimal and most-feasible pathways to implement that system going forward. In so doing, we aim to answer the following research questions:

- Which attributes of an STM system would maximize its potential effectiveness?
- What lessons can be learned from the evolution and operation of international governance in other domains and areas?
- What actions are necessary to kickstart and implement an optimal STM system?

Research Approach

To address these questions and offer relevant policy recommendations, we first surveyed existing literature to identify the leading conceptions of the space traffic problem and any commonly proposed solutions. This review helped inform our understanding of the current legal, political, and technical state-of-play in STM, as well as potential governance structures, previously recommended implementation pathways, and any expected major roadblocks. Then, using both primary sources and secondary analyses, we closely examined the historical development and ongoing execution of global traffic governance in the maritime and air domains. This comparative analysis revealed cross-domain parallels to the present status of global STM, with further implications for possible paths forward in space. In conjunction

---

[9] See Jason Rainbow, "Connecting the Dots: Improving Satellite Collision Predictions for Efficient Space," *SpaceNews*, June 23, 2022 ("Depending on these factors for a given operator, they could perform anywhere from only a few [collision avoidance] maneuvers per year, all the way to multiple maneuvers per week, per satellite,"); Tereza Pultarova, "SpaceX Starlink Satellites Responsible for over Half of Close Encounters in Orbit, Scientist Says," *Space.com*, August 20, 2021 ("an operator managing about 50 satellites will receive up to 300 official conjunction alerts a week").

[10] Generally, *governance* is defined as the ability of ability of an authoritative entity to make and enforce rules and to deliver services. See Francis Fukuyama, "What Is Governance?" *Governance*, Vol. 26, No. 3, July 2013, p. 350. We include in this definition the ability of an authoritative entity to "direct and control the actions or conduct of" entities in the system "either by established laws [rules]" and to "direct and control, rule, or regulate" by agreed upon rules, precedents, and principles. See generally the definition of "govern," *Black's Law Dictionary*, 6th ed., 1990, p. 695.

with general international governance trends, we mined individual success stories of inter-governmental cooperation in other areas for lessons relevant to STM solutions. Synthesizing the conclusions from these research threads yielded preliminary insights and recommendations for realizing effective global STM governance.

In November 2022, to further refine and contextualize these findings, the RAND Corporation hosted three 90-minute virtual workshops with leading STM experts from government, academia, research organizations, and industry. These experts were grouped into one workshop each by the geographical region of the participants' primary affiliations: Europe, Asia-Pacific, and the United States. Each of these regional workshops followed the same agenda with the primary purpose of soliciting feedback on our research approach, conclusions, and, most importantly, our policy recommendations. To ensure diverse perspectives across sectors, participants were selected using a blend of purposive and snowball sampling. Attendance ranged from seven to nine outside participants per workshop; this size facilitated vigorous group discussion. These discussions were held under Chatham House Rules and were not recorded.[11] A RAND team member captured participant input, and real-time opinion surveys were conducted with recorded results (see the appendix). Using these notes and results, we fine-tuned our research and recommendations to reflect the collective expertise of workshop participants.

We acknowledge that our research and analysis in this report is limited to examining potential solutions for STM, particularly as part of our discussion of the various other domains. Because of our focus on solutions that are analogous to space and resource constraints for this work, we do not fully explore many of the aspects (e.g., physical, environmental, technological, etc.) of these domains that make them both different and divergent. These differences represent challenges to adopting the same solutions across domains and, therefore, warrant further study.

Existing Work

Since before the launch of Sputnik, academics have used "traffic" to characterize space activity and to invoke the parallel need for establishing space "rules of the road," or STM. The earliest proposal of such formal regulations came from Lubos Perek at the Twenty-Fifth International Colloquium on Space Law in 1982 with his article "Traffic Rules for Outer Space," which is widely considered to be the origin of STM as a concept and term of art.[12] Thereafter, STM remained a primarily theoretical consideration until 1999 and 2001, when the American Institute of Aeronautics and Astronautics (AIAA) hosted their fifth and sixth Interna-

---

[11] Under Chatham House Rules, information from the workshops can be used by any participants, but the identity of whichever participant made a given comment cannot be disclosed.

[12] Lubos Perek, "Traffic Rules for Outer Space," *Proceedings of the Twenty-Fifth Colloquium on the Law of Outer Space*, AIAA, September–October 1982.

tional Space Cooperation workshops, respectively, which included proceedings dedicated to the topic of space traffic.[13] The second of these workshops culminated in a working group being established at the IAA that was charged with publishing a systematic, interdisciplinary STM report.[14] The resulting *2006 Cosmic Study on Space Traffic Management* was the first comprehensive report in the field of STM and remains a pillar of modern STM literature.[15] The report reinforced the eventual need for international agreement on space traffic rules and conceptualized the necessary components of an effective STM regime. Furthermore, the report recommended initially assigning "operative oversight" over this STM regime, or the "task of space traffic management," to an existing body (e.g., International Civil Aviation Organization [ICAO], UN Office for Outer Space Affairs [UNOOSA]) that would ultimately evolve into a separate, STM-dedicated organization.[16]

Following the 2006 report, heightened academic interest produced research that spanned the various legal, technical, and regulatory dimensions of STM. The resulting patchwork of conferences, workshops, symposia, and independent publications has elevated STM well beyond the purely academic domain (Table 1.1). Multiple national governments have developed or begun developing domestic STM frameworks, and every session of the UNCOPUOS Legal Subcommittee (LSC) since 2016 has included formal deliberations on STM.[17] This proliferation of STM-related activity has solidified a consensus among academics, space operators, and policymakers that the safety and accessibility of future space operations require improvements in traffic governance.[18] However, much debate remains on what form that governance should take and the optimal and most feasible pathway to implementing any future STM regime.[19]

Existing and New Institutions

Prior proposals for improved international STM governance can largely be characterized by how significant of a role they recommend for existing institutions to play in the provision

---

[13] Graham Gibbs and Ian Pryke, "International Cooperation in Space: The AIAA–IAC Workshops," *Space Policy*, Vol. 19, No. 1, February 2003.

[14] Gibbs and Pryke, 2003.

[15] Contant-Jorgenson, Lála, and Schrogl, 2006.

[16] Contant-Jorgenson, Lála, and Schrogl, 2006, p. 15.

[17] Bhavya Lal, Asha Balakrishnan, Becaja M. Caldwell, Reina S. Buenconsejo, and Sara A. Carioscia, *Global Trends in Space Situational Awareness (SSA) and Space Traffic Management (STM)*, Institute for Defense Analysis, 2018.

[18] Hjalte Osborn Frandsen, "Looking for the Rules-of-the-Road of Outer Space: A Search for Basic Traffic Rules in Treaties, Guidelines and Standards," *Journal of Space Safety Engineering*, Vol. 9, No. 2, June 2022; Yun Zhao, "Initial Thoughts on a Possible Regime for Space Traffic Management," *Centre for Aviation and Space Laws*, blog, June 14, 2022.

[19] Paul Larsen, "Space Traffic Management Standards," *Journal of Air Law and Commerce*, Vol. 83, No. 2, 2018.

TABLE 1.1
Sampling of Key Space Traffic Management Activities and Publications

Year	Event	Output	Author or Event Organizer	ISTMO Recommendation	Landmark ISTMO Recommendation
1982	Twenty-Fifth International Colloquium on Space Law	*Traffic Rules for Outer Space*	Lubos Perek	Not applicable	No
1999, 2001	International Space Cooperation workshop	STM working group at IAA	AIAA	Not applicable	No
2006	Publication	2006 *Cosmic Study on Space Traffic Management*	IAA	Operative oversight would be initially administered by existing forum or organization (UNOOSA, ICAO), evolving into separate dedicated STM body	Yes
2007	Publication	*Space Traffic Management*	International Space University (ISU)	Phased system involving the Inter-Agency Space Debris Coordination Committee and UNCOPUOS for rulemaking, credibility, and consensus building, leading to new ICAO-like STM agency	Yes
2007	European Air and Space Conference	STM workshop	Council of European Aerospace Societies (CEAS)	Not applicable	No

Table 1.1—Continued

Year	Event	Output	Author or Event Organizer	ISTMO Recommendation	Landmark ISTMO Recommendation
2008	"Fair and Responsible Use of Space: International Perspective" workshop	"10 Steps to Achieve Fair and Responsible Use of Outer Space" and *The Fair and Responsible Use of Space: An International Perspective* (2010)	European Space Policy Institute, Secure World Foundation (SWF), IAA	Not applicable	No
2011	Publication	*The Need for an Integrated Regulatory Regime for Aviation and Space: ICAO for Space?*	The International Association for the Advancement of Space Safety, European Space Policy Institute	Extend ICA's mandate to space through amending the Chicago Convention or Annexes, creating a "Space Navigation Bureau" (SNB) underneath the ICAO Secretary General	Yes
2014–2019	Space Traffic Management Conferences	STM conference papers	Embry-Riddle Aeronautical University	Not applicable	No
2015	IISL and ECSL Symposium	Influenced UNCOPUOS LSC members to add STM to agenda	International Institute of Space Law (IISL), European Centre for Space Law (ECSL)	Not applicable	No
2015	Colloquium on Law of Outer Space	STM Session	IISL	Not applicable	No
2015–2017	AeroSPACE Symposia	ICAO "Learning Ground" on Civil Space	ICAO, UNOOSA	Not applicable	No
2016–present (as of 2023)	Legal Subcommittee Sessions	Formal deliberations on Space Traffic Management	UNCOPUOS	Not applicable	No

Table 1.1—Continued

Year	Event	Output	Author or Event Organizer	ISTMO Recommendation	Landmark ISTMO Recommendation
2018	Publication	Space Traffic Management: Towards a Roadmap for implementation	IAA	Establish ITU-inspired three-level STM governance regime with governing principles from existing law, traffic rules, and nimble technical standards	Yes
2018	Publication	"Space Traffic Management Standards"	Paul B. Larsen	Use Chicago Convention equivalent or OST protocol to found STM governance structure resembling ICAO, including ANC and variances	Yes
2018	International Astronautical Conference	Founding of Space Traffic Management Committee	International Astronautical Federation (IAF)	Not applicable	No
2020	Publication	"Who Is Right When It Comes to the Right of Way in Space?"	Ruth Stilwell	Developing an STM regime will require international agreement on standards and norms of behavior in space	No
2022	Publication	Cooperative Initiative to Develop Comprehensive Approaches and Proposals for Space Traffic Management (STM)	IAA, IAF, IISL	Create "International Spacefaring Organization," UN-level mandated body comparable to ICAO and IMO	Yes

SOURCES: Lubos, 1983; Gibbs and Pryke, 2003; Contant-Jorgenson, Lála, and Schrogl, 2006; Council of European Aerospace Societies, *1st CEAS European Air and Space Conference*, German Society for Aeronautics and Astronautics, 2007; International Space University, *Space Traffic Management*, 2007; Wolfgang Rathgeber, Kai-Uwe Schrogl, and Ray Williamson, eds., *The Fair and Responsible Use of Space: An International Perspective*, Springer, 2010; Ram S. Jakhu, Tommaso Sgobba, and Paul Stephen Dempsey, eds. *The Need for an Integrated Regulatory Regime for Aviation and Space: ICAO for Space?* Springer, 2011; Space Traffic Management Conference 2019: Progress Through Collaboration, Embry-Riddle Aeronautical University, February 26–27, 2019; Space Law Symposium 2015, International Institute of Space Law and European Centre for Space Law, April 13, 2015; ICAO/UNOOSA AeroSPACE Symposium, Montréal, Canada, March 18–20, 2015; United Nations Committee on the Peaceful Uses of Outer Space, Legal Subcommittee , "2016 LSC Draft Report," United Nations Office for Outer Space Affairs, April 7, 2016; Kai-Uwe Schrogl, Corinne Jorgenson, Jana Robinson, and Alexander Soucek, eds., *Space Traffic Management—Towards a Roadmap for Implementation*, International Academy of Astronautics, 2018; Larsen, 2018; Ruth Stilwell, "Who Is Right When It Comes to the Right of Way in Space?" paper presented at Facing the Security Challenge, 6th Annual Space Traffic Management Conference, University of Texas, February 19–20, 2020; International Academy of Astronautics, International Astronautical Federation, and International Institute of Space Law, 2022.

NOTE: ITU = International Telecommunication Union. ANC = Air Navigation Commission.

of "operative oversight" of STM. Some proposals argue that expanding current organizations would be sufficient, others contend that a new body is needed, and some recommend a blend or sequence of governance structures that involve existing and new entities, such as the phased proposal in Contant-Jorgenson, Lála, and Schrogl (2006).

The most comprehensive example of a model that expands existing institutions is a landmark 2011 study by the International Association for the Advancement of Space Safety and European Space Policy Institute that outlines an STM regime and proposes that the aforementioned operative oversight be provided by ICAO through an extension of its mandate to oversee space (up to and including geostationary orbit) (Figure 1.1).[20] By amending the foundational Chicago Convention and Annexes to establish an SNB under the ICAO Secretary General and parallel to the Air Navigation Bureau and adopting the existing "high seas" framework, ICAO could provide the necessary oversight for managing space traffic. This proposed SNB solution leverages ICAO's existing credibility and space-related activity but has considerable limitations related to ICAO's ability to simultaneously address the legal, technical, and regulatory challenges that are specific to the air and space domains.[21] However, some level of integration between air and space remains a priority for many STM experts.[22] Other scholars have suggested the inverse approach of extending the purview of existing space organizations (e.g., UNOOSA, Inter-Agency Space Debris Coordination Com-

---

[20] Jakhu, Sgobba, and Dempsey, 2011. While the ICAO Convention, article 3, notes that ICAO governs international civil aviation and is not applicable to *state aircraft*, which are generally considered to be aircraft used in national military, customs, and police services, military aircraft generally follow ICAO rules as a matter of directive or in the normal course of operations. See, for example, Department of Defense Instruction 4540.1, *Use of International Airspace by U.S. Military Aircraft and for Missile and Projectile Firings*, change 1, May 22, 2017. Given the national military presence in space, it would be difficult, if not impossible, to have an effective STM organization that did not similarly consider the management of state space objects.

[21] ICAO/UNOOSA AeroSPACE Symposium, 2015; Stephen Hunter, "How to Reach an International Civil Aviation Organization Role in Space Traffic Management," paper presented at the Space Traffic Management Conference 2014: Roadmap to the Stars, Daytona Beach, Fla., November 5–6, 2014, p. 17; Ruth E. Stilwell, Diane Howard, and Sven Kaltenhauser, "Overcoming Sovereignty for Space Traffic Management," *Journal of Space Safety Engineering*, Vol. 7, No. 2, June 2020; Yu Takeuchi, "STM in the Nature of International Space Law," paper presented at Space Traffic Management Conference 2019: Progress Through Collaboration, Daytona Beach, Fla., February 26–27, 2019, p. 9; Additionally, note that ICAO's mandate is restricted to civil aviation. Extending ICAO to space without addressing military operations might be inadequate in ensuring safety in orbit, especially as the number of military satellites in space continues growing, and entanglement, or dual use, becomes more popular. UNOOSA and UNCOPUOS are not limited in the same way.

[22] Michael Chatzipanagiotis, "Looking into the Future: The Case for an Integrated Aerospace Traffic Management," paper presented at the 58th IISL Colloquium on the Law of Outer Space, Jerusalem, Israel, October 2015; Martin Griffin, "Integration of Aerospace Operations into the Global Air Traffic Management System," paper presented at the Space Traffic Management Conference 2014: Roadmap to the Stars, Daytona Beach, Fla., November 5–6, 2014, p. 12; Sanat Kaul, "Integrating Air and Near Space Traffic Management for Aviation and Near Space," *Journal of Space Safety Engineering*, Vol. 6, No. 2, June 2019.

FIGURE 1.1

Example Space Traffic Management Organization Structure as Incorporated into International Civil Aviation Organization

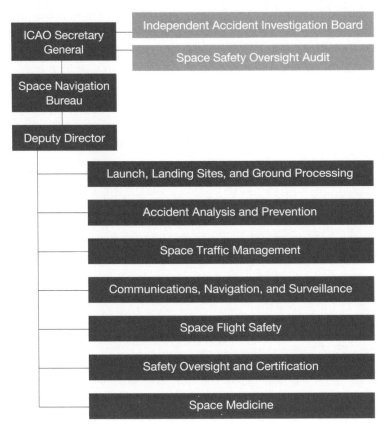

SOURCE: Adapted from Jakhu, Sgobba, and Dempsey, 2011, p. 133.

mittee) into traffic management (as opposed to extending traffic organizations into space) but have not described detailed implementation pathways or structures.[23]

The other most common proposal for centralized STM involves establishing a new STM-dedicated body but, typically, modeling that body after intergovernmental examples from other domains, such as air, maritime, and telecommunications (Figure 1.2). The International Space University (ISU) published the earliest institutional analysis of STM-capable candidates and ultimately recommended that an international space management organization be founded with the narrow mandate of STM, outside the UN but with early involve-

---

[23] Kai-Uwe Schrogl, "Space Traffic Management: The New Comprehensive Approach for Regulating the Use of Outer Space—Results from the 2006 IAA Cosmic Study," *Acta Astronautica*, Vol. 62, Nos. 2–3, January–February 2008.

FIGURE 1.2

Example Space Traffic Management Organization Structure as a Separate Entity

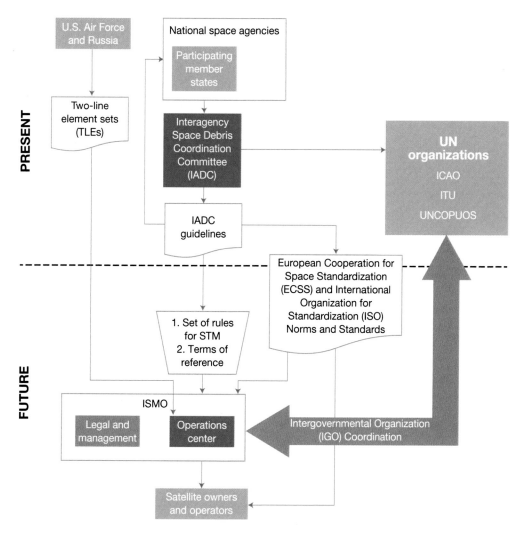

SOURCE: Adapted from International Space University, 2007, p. 53.

ment from the Inter-Agency Space Debris Coordination Committee and continued coordination with ICAO, ITU, and UNCOPUOS.[24] Paul Larsen's seminal 2018 work modernized and expanded the ISU analysis with a systematic comparison of existing institutional models, including a nongovernmental, private standard-setting organization such as the American

---

[24] ISU, 2007.

Bureau of Shipping.[25] Larsen concluded that space-applicable elements of ICAO and its ANC provide the most promising prototype for a future independent civilian international STM body, which would be established through a Chicago Convention equivalent or Outer Space Treaty protocol.

The IAA's milestone 2018 update to its 2006 report endorsed a three-level STM organization that would mirror the ITU with a foundation of governing principles rooted in existing space law, a substructure of robust space traffic rules, and a catalogue of more-flexible technical standards.[26] In 2022, a comprehensive, multiyear joint IAA-IISL-IAF STM working group culminated in a report that recommended an "International Spacefaring Organization," a new UN-level International Maritime Organization (IMO) and International Civil Aviation Organization (ICAO) equivalent for STM.[27] Although an STM-dedicated organization would offer more flexibility than do existing institutions to address unique space traffic challenges, credibility and buy-in remain critical hurdles to the establishment of any new intergovernmental body.

Top-Down and Bottom-Up Approaches

Independent of structure, scholars agree that any centralized international STM regime would constitute a major departure (or "big bang") from the current state of play, which is mostly characterized by the piecemeal development of nonbinding national norms and frameworks.[28] Although Russia and China have supported new STM rulemaking at the international level, the United States, which has the most advanced national STM regime, has a vested and stated interest in maintaining the voluntary and federated direction of the status quo, deadlocking STM negotiations at UNCOPUOS LSC sessions.[29]

---

[25] Larsen, 2018.

[26] Schrogl et al., 2018.

[27] International Institute of Space Law, "IISL, IAA and IAF Conclude Major Report on STM—International Institute of Space Law," undated.

[28] Bryon C. Brittingham, "Does the World Really Need New Space Law," *Oregon Review of International Law*, Vol. 12, No. 1, 2010; Kai-Uwe Schrogl, "Space Traffic Management," No. 3, 2007, p. 4; Theresa Hitchens, "Forwarding Multilateral Space Governance: Next Steps for the International Community," working paper, Center for International and Security Studies at Maryland, University of Maryland, August 2018, p. 38; Lal et al., 2018.

[29] Travis S. Cottom, "Creating a Space Traffic Management System and Potential Geopolitical Opportunities," *Astropolitics*, Vol. 19, No. 1–2, May–August 2021; Michael P. Gleason, "Establishing Space Traffic Management Standards, Guidelines and Best Practices," *Journal of Space Safety Engineering*, Vol. 7, No. 3, September 2020; Hitchens, 2018. We note, however, that recent statements by some U.S. officials indicate a softening to the possibility of developing binding international rules with respect to space. These statements have been made in support of the U.N. open-ended working group on "reducing space threats through norms, rules, and principles of responsible behaviors." Although the statement does not specifically mention developing binding rules for STM, collision avoidance is a topic for the OEWG. Eric Desautels, "U.S.

This geopolitical reality has framed an ongoing debate within contemporary STM scholarship between a top-down or bottom-up pathway to establishing international space traffic rules. A comparison between the two pathways is shown in Figure 1.3. Proponents of the *top-down* approach support establishing an intergovernmental STM body (such as those detailed in Figures 1.1 and 1.2) first, without waiting for individual nations to develop domestic STM regimes. Proponents of this pathway contend that this approach ensures multilateral participation from the start, accelerates the global rulemaking process, and avoids incoherence or fragmentation at the international level.

Alternatively, proponents of a *bottom-up* approach argue that a critical mass of strong national STM frameworks must be developed first before harmonizing those frameworks internationally, which would mirror how ICAO evolved from preexisting domestic air traffic regimes.[30] This pathway would provide nations with near-term flexibility to make timely responses to domestic STM issues, but it also risks fragmentation and delays international agreement until a sufficient number of states develop their own STM capabilities.[31] Some literature on the bottom-up approach still envisions the eventual creation of an intergovern-

FIGURE 1.3.
Bottom-Up Versus Top-Down Implementation Comparison

Bottom-up implementation

STM → States

First: States each develop national frameworks.

Then: States harmonize frameworks into a global regime with an oversight body.

Pros	Mirrors the proven evolution of ICAO and IMO
	Allows for near-term flexibility to address national concerns
Cons	Risks fragmentation and Russia or China nonparticipation
	Delays urgently needed global standards until enough states develop own frameworks
Example	Organize UN harmonizing convention after a sufficient number of national STM regimes are established

Top-down implementation

STM → States

First: States negotiate the global regime and establish an oversight body.

Then: States implement and adopt the new STM rules.

Pros	Prioritizes multilateral participation from the start of accelerated rule-making process
	Avoids international incoherence
Cons	Could postpone domestic reforms and face steep industry pushback
	Few national frameworks to harmonize now, limiting input and implementation expertise
Example	Amend Chicago Convention to extend ICAO to space and establish global STM rules

Statement to the Open Ended Working Group on Reducing Space Threats Through Norms, Rules and Principles of Responsible Behavior," U.S. Mission to International Organizations in Geneva, May 9, 2022.

[30] Brian Weeden, "Muddling Through Space Traffic Management," *SpaceNews*, September 22, 2017.

[31] Schrogl et al., 2018.

mental STM organization to oversee the harmonized regime only after national frameworks are widely established. Other experts, however, believe the modern geopolitics of space will restrict possible international outcomes of the bottom-up pathway to simple "coordination," instead of "management," focused only on space situational awareness (SSA) data-sharing improvements without comprehensive STM rules or a centralized authority.[32] Regardless, experts agree that either the top-down or bottom-up implementation pathway toward safer space traffic will need to overcome civil-military, public-private, sovereignty, technological, and political challenges, requiring further multidisciplinary research in the still-nascent field of STM.[33]

Building on the ongoing debates about the optimal model and pathway to a centralized STM system, the following chapters examine possible insights and transferrable lessons from traffic management in the maritime and air domains.

---

[32] Theodore J. Muelhaupt, Marlon E. Sorge, Jamie Morin, and Robert S. Wilson, "Space Traffic Management in the New Space Era," *Journal of Space Safety Engineering*, Vol. 6, No. 2, June 2019; Daniel L. Oltrogge, "The 'We' Approach to Space Traffic Management," paper presented at the 2018 SpaceOps Conference, American Institute of Aeronautics and Astronautics, May 28–June 1, 2018.

[33] Ntorina Antoni, Christina Giannopapa, and Kai-Uwe Schrogl, "Legal and Policy Perspectives on Civil–Military Cooperation for the Establishment of Space Traffic Management," *Space Policy*, Vol. 53, August 2020; Frandsen, 2022; Stilwell, Howard, and Kaltenhauser, 2020.

CHAPTER 2

Coordination in the Maritime Domain

With its large swaths of international waters and the presence of vessels from across the globe, the maritime domain naturally lends itself to comparisons to space. Indeed, many theorists of space power and space law have drawn explicit analogies to humanity's use of the sea.[1] Although the maritime domain's history spans millennia, this chapter outlines how the domain's governance structures developed more recently in a clear pattern. First, industry and individual states began to create norms and some binding bilateral rules. These norms and initial rules then developed further, albeit slowly and with much debate, to inform international law. This development led to the eventual creation of treaties and a specialized UN agency overseeing their implementation. Finally, new technologies continue to play a facilitating role in moving states toward international and nearly universal adoption of the governance scheme.

This chapter and Chapter 3 expand on recent RAND analysis of the maritime and air domains' governance mechanisms and their applicability to the space domain.[2] In the maritime sector, regulations for preventing collisions at sea have achieved widespread adoption. We note that although such aspects of maritime governance as those affecting environmental conservation, pollution mitigation, and security have relevance to comparable issues in space, they are outside the scope of this STM-focused report.

Early Developments of Maritime Norms and Rule-Building

Freedom of the Seas and Coastal Waters

Although early efforts to create maritime governance were made by states, some of the earliest norms and rules in sea travel and trade came from prominent jurists. For many centuries, kingdoms and states proclaimed territorial control over vast tracts of ocean even if they were

---

[1] See Julian Corbett, *The League of Nations and Freedom of the Seas*, Oxford University Press, London, 1918; Hugo Grotius, *Mare Liberum [The Free Sea]*, trans. by Richard Hakluyt, Liberty Fund, [1609] 2004; Alfred Thayer Mahan, *The Influence of Sea Power Upon History 1660–1783*, 12th ed., Project Gutenberg, [1890] 2004.

[2] Dan McCormick, Douglas C. Ligor, and Bruce McClintock, *Cross-Domain Lessons for Space Traffic: An Analysis of Air and Maritime Treaty Governance Mechanisms*, RAND Corporation, RR-A2208-2, 2023.

limited in their means to enforce those claims in an age of sail. One of the defining examples of such claims came in the 1494 Treaty of Tordesillas, in which Spain and Portugal attempted to establish and divide sovereign control over the Americas and their adjacent seas.[3] These claims led to pushback from competing states, creating the basis for a modern movement that favored freedom of the seas. Perhaps the most-prominent work on maritime norms came from Dutch legal scholar Hugo Grotius in *Mare Liberum*, where he argued that the high seas, by their nature, must be free to navigate.[4] Grotius was not the first to argue for freedom of the seas, but he gained traction by couching his argument in a sense of natural law and justice and noting that the seas were not subject to occupation or conquest.[5] In debating this claim, other jurists and state leaders argued for some degree of control over coastal waters, leading to one of the first widely accepted maritime norms in Europe by the 1700s. The *cannon-shot rule* held that coastal states enjoyed sovereignty over the seas adjacent to their shores out to the distance of a cannon's range.[6]

By the middle of the 19th century, despite continued debate about the extent of coastal sovereignty, freedom of the seas had developed into a widely accepted norm that governed travel on the high seas, and several major seafaring powers began attempting to formalize this norm into laws.[7] At the conclusion of the Crimean War, seven European participants signed the 1856 Paris Declaration Respecting Maritime Law, which reaffirmed that privateering "was and remained abolished."[8] It also dictated that vessels flying neutral flags and the goods onboard those vessels were not liable to capture except for "war contraband."[9] The treaty emphasized that the signatories would make efforts to invite nonparticipant states to accede to the terms.[10] Other instances of multilateral and bilateral rulemaking to guarantee freedom of movement on the high seas followed the Paris Declaration, often initiated by suggestions from private organizations.[11]

---

[3]   National Geographic, "Jun 7, 1494 CE: Treaty of Tordesillas," updated October 4, 2022.

[4]   Grotius argued that the states could not physically occupy any part of the oceans. Tullio Treves, "Historical Development of the Law of the Sea," in Donald Rothwell, Alex Oude Elferink, Karen Scott, and Tim Stephens, eds., *The Oxford Handbook of the Law of the Sea,* Oxford University Press, 2015, pp. 1–2.

[5]   George P. Smith II, "The Politics of Lawmaking: Problems in International Maritime Regulation—Innocent Passage v. Free Transit," *University of Pittsburgh Law Review,* Vol. 37, No. 3, 1976, pp. 487–489.

[6]   Treves, 2015.

[7]   Treves, 2015, p. 4.

[8]   The seven states were Austria, France, Prussia, Russia, Sardinia, Turkey, and the United Kingdom. Paris Declaration Respecting Maritime Law, April 16, 1856; Marie Jacobsson, "Institutional Arrangements for the Ocean: From Zero to Indefinite?" *Ecology Law Quarterly,* Vol. 46, No. 1, March 31, 2019, p. 306.

[9]   Stockton, Charles H., "The Declaration of Paris," *American Journal of International Law,* Vol. 14, No. 3, July 1920, pp. 356–68.

[10]   Paris Declaration Respecting Maritime Law, 1856.

[11]   Jacobsson, 2019, p. 306.

First Attempts at International Governance for Navigation

As interactions between vessels from around the world increased, navigation safety presented a growing challenge on an international scale. Prior to the late 1800s, no specific or codified rules governed vessels' approaches in open water.[12] As steam-powered vessels became common, and the number of collisions between vessels rose, states had more incentives to unify rules of the road to avoid those accidents.[13] In 1863, Britain and France, as the leading maritime powers of the time, made an initial effort to solve this problem by signing an agreement to abide by identical regulations governing vessel interaction at sea.[14] The agreement led dozens of other states to follow suit and adopt the same or slightly altered rules.[15] In 1864, amid a civil war marked by a controversial blockade of maritime trade with the southern states, the United States adopted similar language and added provisions on overtaking vessels and the use of sound signals on approach.[16]

A more significant effort at achieving uniformity came in 1889 at the first International Maritime Conference in Washington, D.C. The United States convened the meeting, and 26 other nations attended.[17] U.S. legislators drafted versions of the rules that were proposed in the meeting, but it took nearly a decade of revisions until the U.S. Congress finally reached agreement as to certain, but not all, rules, which were then formally enacted into law in 1897.[18] Despite this progress, many states still operated under their own regulations unmoored to a uniform, global system of rules governing sea travel and maneuver.

Two final instances of maritime rulemaking informed the future of international governance in the domain. One major international effort at formalizing maritime rules came at the League of Nations 1930 Codification Conference in The Hague ("Hague Conference").[19] Ambitiously, the League sought to fully codify the law of the sea, but no major agreements came from the event as the parties strongly disagreed over the breadth of territorial seas and fishing rights.[20] Following the failure of the Hague Conference, some states returned to unilateral rulemaking. For example, the United States issued the Truman Proclamations in 1945, claiming sovereignty over its continental shelf and the resources therein, far outstripping the

---

[12] Winford W. Barrow, "Consideration of the New International Rules for Preventing Collisions at Sea," *Tulane Law Review*, Vol. 51, No. 4, 1976–1977, p. 1182.

[13] Melvin J. Tublin, "The New Danger Signal Authorized by the International Rules of the Road," *Georgetown Law Journal*, Vol. 42, No. 1, November 1953, pp. 95–96.

[14] Tublin, 1953, p. 96.

[15] Tublin, 1953.

[16] Jeff Werner, "The History of the Rule of the Road—Sailing Vessel History," *All at Sea*, January 26, 2017.

[17] Tublin, 1953, p. 96.

[18] Tublin, 1953.

[19] Daniel Patrick O'Connell, "The History of the Law of the Sea," in Ivan Anthony Shearer, ed., *The International Law of the Sea*, Vol. 1, 1st ed., Clarendon Press, 1982, p. 20.

[20] O'Connell, 1982.

previous norms of the cannon-shot rule; several other states responded with similar claims over 200 nautical miles of sea adjacent to their shorelines.[21]

The sudden growth of unilateral claims and the new enthusiasm for multilateralism in the wake of the World War II likely led the global community toward a more serious effort at codification. The first session of the UN International Law Commission occurred in 1949.[22] The International Law Commission went through various drafts for nearly a decade before submitting a final report to the UN General Assembly. The report served as the basis for four conventions that were opened for signature at the first UN Conference on the Law of the Sea (UNCLOS) in 1958.[23]

International Mechanisms: United Nations Conference on the Law of the Sea and the International Maritime Organization

As of 2023, a well-developed set of international laws and institutions govern the maritime domain, and participating states serve as key enforcement mechanisms. The negotiations surrounding UNCLOS in the latter half of the 20th century have involved several procedural hurdles. Awareness of how these hurdles have been handled could inform a similar process in building a space governance regime, and the development of IMO could likewise provide a blueprint for an ISTMO. This section outlines the procedural roadblocks that states overcame to build a maritime governance regime and summarizes the form and function of both UNCLOS and IMO.

Procedural Issues

Both UNCLOS and IMO were delayed because of procedural burdens within the UN. The 1948 UN conference in Geneva adopted a convention creating the Governmental Maritime Consultative Organization (which would later become IMO), but the convention did not enter into force until 1958 because the members could not meet the 21-vote threshold required for adoption.[24] Continued disagreement over territorial seas prevented the 86 attending states at UNCLOS I from reaching the required two-thirds of votes to achieve agreement on the distance from the shoreline that would constitute the territorial sea.[25] The members did break

---

[21] Rodrigo Facalossi de Moraes, "The Parting of the Seas: Norms, Material Power, and State Control over the Ocean," *Revista Brasileira de Política Internacional* [*Brasilian Review of International Politics*], Vol. 62, No. 1, April 15, 2019.

[22] de Moraes, 2019.

[23] Treves, 2015, p. 11.

[24] Delays largely resulted from states' concerns about potential economic regulation coming within IMO's purview. IMO, "Convention on the International Maritime Organization," webpage, undated-c.

[25] O'Connell, 1982, pp. 24–25.

the UNCLOS I proposals into four conventions and an optional protocol to try to ease the passage of those conventions, but members still could not achieve consensus on the role of compulsory settlement of disputes or the width of territorial seas.[26] In 1960, members at UNCLOS II also failed to meet the two-thirds threshold on the issue of exclusive fishery rights, leading to another two decades of negotiations and piecemeal unilateral legislation by some states.[27]

During UNCLOS III, from 1973 to 1982, participants made procedural adjustments that helped move the debate forward. The general assembly broke the work into committees that covered different topics and advised each committee to produce a negotiating text essentially summarizing their internal debate. The committees created a revised text based on these debates that remained nonbinding, but any states that sought to further amend the text needed to build a large coalition to meet committee thresholds for amendment. Through this method of multiparty negotiation and the balancing of trade-offs, the members built consensus throughout the following decade and adopted a convention on the Law of the Sea in 1982.[28] Even then, because of disagreements related to deep seabed mining, the law did not enter into force until 1994, when it received the requisite 60 signatures. UNCLOS III has been signed by 167 parties, including the European Union (EU). Despite its influential role in the negotiations, the United States has signed but not ratified the agreement, making it one of the last major holdouts to UNCLOS III.[29]

Administrative and Legal Structure

As UNCLOS negotiations progressed, IMO developed into the sole specialized UN agency that was dedicated to building global standards on safety, security, and environmental matters for the international maritime domain.[30] IMO implements various conventions on maritime law, ensures uniformity across the domain, and regularly adopts updates to relevant maritime conventions, such as the Convention on the Safety of Life at Sea (SOLAS) and the Convention on the International Regulations for Preventing Collisions at Sea (COLREGs).[31] However, similar to other UN agencies, IMO lacks direct enforcement authority. Instead,

---

[26] Tulio Treves, "1958 Geneva Conventions on the Law of the Sea," Audiovisual Library of International Law, September 2008.

[27] de Moraes, 2019; O'Connell, 1982, p. 26.

[28] O'Connell, 1982, pp. 26–29.

[29] Anya Wahal, "On International Treaties, the United States Refuses to Play Ball," *The Internationalist*, blog, Council on Foreign Relations, January 7, 2022. See also P. Hoagland, J. Jacoby, and M. E. Schumacher, "Law of the Sea," in John H. Steele, ed., *Encyclopedia of Ocean Sciences*, 2nd ed., Elsevier 2001, p. 1481.

[30] IMO, "Introduction to IMO," webpage, undated-g.

[31] IMO, "Convention on the International Regulations for Preventing Collisions at Sea, 1972 (COLREGs)," webpage, undated-d; IMO, "International Convention for the Safety of Life at Sea (SOLAS), 1974," webpage, undated-f; IMO, "Structure of IMO," webpage, undated-i.

IMO relies on its 175 member states to incorporate regulations into their national frameworks and enforce standards in operating ports and vessels.[32]

IMO has an assembly of all members that controls budgets and elects the Council, which consists of a rotating group of 40 states (the structure of IMO is shown in Figure 2.1).[33] The Council is composed of members from three separate categories: (1) ten states with the largest interest in providing shipping services, (2) ten states with the largest interest in international maritime trade, and (3) 20 states that "have special interests in maritime transport or navigation and whose election to the council will ensure the representation of all major geographic areas of the world."[34] Headquartered in London, IMO also employs approximately 300 civil servants in its Secretariat to maintain domain expertise and conduct daily operations.[35]

Finally, numerous committees oversee different subject-matter areas, the most important of which is the Maritime Safety Committee.[36] The Maritime Safety Committee oversees the

FIGURE 2.1.
International Maritime Organization Structure

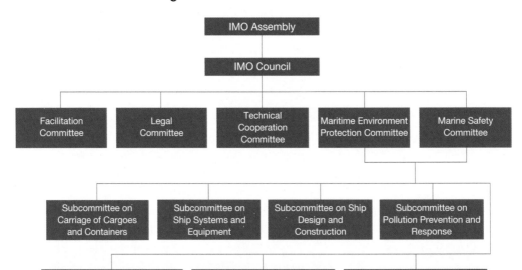

SOURCE: Adapted from United Arab Emirates Ministry of Energy & Infrastructure, "International Maritime Organization," webpage, undated.

---

[32] IMO, "Frequently Asked Questions," webpage, undated-d.

[33] IMO, undated-i.

[34] IMO, undated-i.

[35] U.S. Coast Guard, "USCG IMO Homepage," webpage, updated February 6, 2018.

[36] IMO, "Safety of Navigation," webpage, undated-h.

framework for navigation safety found in the SOLAS Convention and COLREGs, approves all technical requirements for vessels, and has the authority to adopt and amend traffic separation schemes in crowded waterways.[37] The convention that created IMO left the question of funding for the assembly to decide, leading to a unique arrangement in which each member state pays a contribution based on its overall contribution to the UN and its registered tonnage of merchant shipping.[38]

United Nations Conference on the Law of the Sea and International Maritime Organization Navigation Governance

IMO oversees and adjusts technical rules, and the provisions of UNCLOS delineate different zones of international and state-controlled waters and govern how vessels may operate in those zones. Under UNCLOS, oceans are divided into seven zones: (1) the territorial sea, (2) the contiguous zone, (3) the exclusive economic zone, (4) the continental shelf, (5) the extended continental shelf, (6) the high seas, and (7) the deep seabed area (Figure 2.2).[39] Each of these zones is accompanied by a unique set of regulations, but the high seas and the area are both considered a common heritage of humankind, and thus only a limited legal framework applies in those two zones.[40]

In addition to establishing maritime zones, "UNCLOS codifies the doctrine of flag state primacy" as a key enforcement mechanism of maritime law.[41] For centuries, flags have been used to identify the origin of a ship. Dating back to Ancient Rome, flags of convenience, which show a different state of registration to the ship's actual registration, have been used to avoid regulations and financial charges.[42] Because of the frequency of use of flags of convenience, Article 94 of UNCLOS III codifies the duties of a flag state. According to UNCLOS III, once a "flag state" extends its nationality upon a ship, that ship has the nationality of the state whose flag it is flying, as opposed to the nationality of the ship's owner, operator, or crew.[43] To ensure free navigation in the high seas, nations must have their ships flagged; otherwise, the ships are considered stateless and under the jurisdiction of other nations. Once flagged, ships follow the regulations laid out in UNCLOS III: (1) states assume jurisdiction

---

[37] IMO, *Status of Multilateral Conventions and Instruments in Respect of Which the International Maritime Organization or Its Secretary-General Performs Depositary or Other Functions*, July 31, 2013.

[38] IMO, "Convention on the International Maritime Organization," webpage, undated-c.

[39] UNCLOS, 1982.

[40] UNCLOS III, Part XI, Article 136, 1982.

[41] Andrew J. Norris, "The 'Other' Law of the Sea," *Naval War College Review*, Vol. 64, No. 3, 2011, p. 80.

[42] HG Legal Resources, "What Is a Flag of Convenience?" webpage, undated.

[43] Norris, 2011, p. 80.

FIGURE 2.2

UN Convention on the Law of the Sea Ocean Zones

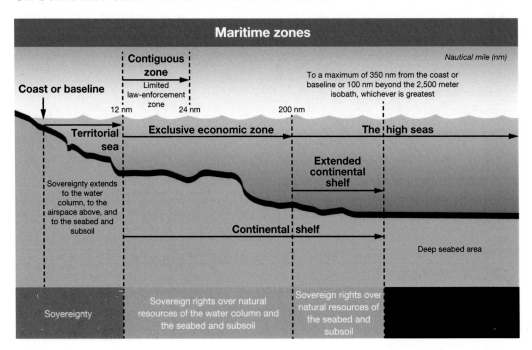

SOURCE: Adapted from Margaret Bohan, "NOAA's Participation in the U.S. Extended Continental Shelf Project," National Oceanic and Atmospheric Administration Office of Ocean Exploration and Research, undated.

over each ship flying its flag and apply internal law to those aboard, and (2) states take measures to ensure safety at sea.[44]

When UNCLOS entered into force in 1994, the first meeting of States Parties to the Law of the Sea convened in New York. Since then, the foundation of UNCLOS has remained untouched, largely because it is written in general terms with opportunities for interpretation and addition via other mechanisms. As of 2023, UNCLOS signatories meet annually to elect the members of the International Tribunal for the Law of the Sea (ITLOS). One-third of members of the tribunal are elected every three years. ITLOS handles disputes that arise out of the interpretation and application of UNCLOS.[45] In addition to ITLOS, Part XV of UNCLOS provides three additional dispute settlement mechanisms for signatories, which include the International Court of Justice, an arbitral tribunal in accordance with Annex VII

---

[44] UNCLOS, 1982. We note that the current practice of flag states pursuant to UNCLOS is not without its critics, who warn that the practice has incentivized a "race to the bottom" in terms of regulation and compliance as states seek to register with countries that have less stringent legal and regulatory schemes. See Allan I. Mendelsohn, "Flags of Convenience: Maritime and Aviation," *Journal of Air Law and Commerce*, Vol. 79, No. 1, Winter 2014, pp. 153–154.

[45] International Tribunal for the Law of the Sea, "Latest News," webpage, undated.

of UNCLOS, and a special arbitral tribunal in accordance with Annex VIII of UNCLOS.[46] Regardless of which mechanism a state chooses, dispute resolution is compulsory under Part XV of UNCLOS.[47]

During these annual meetings, States Parties also elect all 21 members of the Commission on the Limits of the Continental Shelf (CLCS) every five years. They also consider the report of ITLOS and address various budgetary and administrative issues. The Secretary-General of the International Seabed Authority and the Chairman of the CLCS provide general information during the annual meeting and address issues that have arisen in relation to UNCLOS.[48]

The United States has not officially signed UNCLOS III, but it has adhered to the principles of UNCLOS as a matter of binding customary international law and has served as an enforcement mechanism for its signatories. However, UNCLOS III is binding on signatory states only, which means that nonsignatory states, such as the United States, are not required to comply in terms of the treaty and its specific provisions. The majority of nations, even nonsignatory nations, abide by the provisions in UNCLOS III and regard it as customary international law, similar to the United States.[49]

Technological Innovations

Technological development has further facilitated the movement toward international maritime governance and uniformity of rules across jurisdictions. Two prominent examples of technology that drive global maritime governance are onboard sensors and automation. These systems function together to provide the information that vessels need to avoid collisions and otherwise safely and efficiently navigate.

An automatic identification system (AIS) is an example of an automated onboard sensor. An AIS communicates self-identification and positioning information on vessels and shore stations.[50] By allowing the marine community to track vessels, "port authorities and maritime safety bodies can manage maritime traffic and reduce hazards of marine navigation."[51] An AIS uses very high frequency transponders to communicate position information based on the Global Positioning System or other internal sensors. The transmitted information identifies the vessel, the vessel's position and movement, and the navigational information of nearby AIS-equipped vessels. Global adoption of AIS simplifies information exchange

---

[46] International Relations and Defence Committee, *UNCLOS: The Law of the Sea in the 21st Century*, House of Lords, March 1, 2022.

[47] UNCLOS, 1982, Part XV.

[48] UN, "Meeting of States Parties to the 1982 United Nations Convention on the Law of the Sea," webpage, March 15, 2022.

[49] Norris, 2011, p. 85.

[50] NATO Shipping Centre, "AIS (Automatic Identification System) Overview," NATO, 2021.

[51] NATO Shipping Centre, 2021.

by enabling efficient and timely communication between vessels and enhances situational awareness in the maritime domain.[52]

IMO added requirements for such ship location reporting systems to the SOLAS convention in 1994. As of another SOLAS convention update in 2000, the IMO Maritime Safety Committee requires that all ships of 300 gross tonnage and upward, cargo ships of 500 gross tonnage and upward, and all passenger ships broadcast their position with AIS, which allows vessels to be trackable even in the most-remote areas.[53] According to *Global Fishing Watch*, over 400,000 AIS devices broadcast vessel information each year. However, this excludes many fishing vessels, which are only required to be equipped with AIS if they are longer than 65 feet and might have incentives to mask their movements when fishing in certain waters. Still, the increasing presence of AIS demonstrates that international governance mechanisms can facilitate the adoption of new technology to ensure safer traffic management.

Technological innovations could lead to further IMO regulations or to other legal developments in the future, especially in the wake of the global coronavirus disease 2019 (COVID-19) pandemic. The international community recognized the importance of automation and remote operation in maritime transport to the ability to withstand such a disruption as a global pandemic. One example of these technological advances is the smart ship. These ships are "equipped with digital tools to improve vessel performance and maintenance, as well as to enable remote monitoring and real-time support of vessel operations."[54] Efforts are also being made by IMO to develop maritime autonomous surface ships (MASS), drones, and navigation systems. As trials continue, MASS will initially be used for autonomous shipping around the world and will be operated from an onshore Remote Operating Centre.[55] Unmanned vessels also have the potential to collect ocean data and detect surface and subsurface threats. The IMO Maritime Safety Committee has been developing a nonmandatory goals-based MASS code for cargo ships that it hopes to adopt in 2024.[56] A mandatory MASS code is expected to follow in 2028, and the Convention on Facilitation of International Maritime Traffic will likely be the main facilitator of these standards and recommended practices.[57] MASS, like AIS, could be another example of technology that requires such institutions as IMO to enable widespread adoption to make the domain as a whole safer and more efficient.

---

[52] Stephen Garber and Marissa Herron, "How Has Traffic Been Managed in the Sky, on Waterways, and on the Road? Comparisons for Space Situational Awareness (Part 2)," *Space Review*, June 15, 2020.

[53] Global Fishing Watch, "What Is the Automatic Identification System (AIS)?" webpage, undated; International Maritime Organization, "AIS Transponders," webpage, undated-a.

[54] Julian Clark, ed., *Shipping Laws and Regulations 2022–2023*, Global Legal Group, 2022.

[55] Paul Dean, Tom Walters, Jonathon Goulding, and Henry Clack, *Autonomous Ships: MASS for the MASSes*, Holman Fenwick Willan, 2022

[56] IMO, "Autonomous Shipping," webpage, undated-b.

[57] IMO, undated-b.

Conclusion

The long history of the maritime domain contains several lessons that are applicable to space governance. Pressure to generate international governance schemes came from both industry and states themselves as several early attempts at regional and international schemes came up short. Despite early failures, the existence of long-developed norms eased the transition to international governance. Moreover, the maritime domain demonstrates the need for the representation of various interests within an agency through apportioning funding fairly and rotating membership of voting bodies to account for states of various geographic areas and access to the domain. Finally, maintaining technical expertise at the international level serves to facilitate sound rulemaking and adoption of new technologies, which can in turn assist with boosting domain awareness and safety.

CHAPTER 3

Coordination in the Air Domain

The exploitation of the air domain has a much shorter history than in the maritime sector. In many ways governance of the airways more readily lends itself to comparisons to space.[1] The shorter timeline from first flight to international governance than in the maritime domain and the pattern of steadily increasing access to the air domain stand out as prominent similarities between the air and space domains.

History of the Development of the Air Domain

Since the birth of fixed-wing civil aviation in 1903, the nature of air travel made enforcement of local ordinances difficult and quickly led many operators to see a need for national, and eventually international, governance. As commercial and private aviation operations increased, concern for the safety of both the passengers in the aircraft and the public on the ground led to the first attempts at regulation. In the United States, Connecticut passed the first state air law (based on an automobile law) in 1911, which required registration of aircraft and a pilot's license.[2] Massachusetts and Hawaii followed with similar rules, but cross-country flights demonstrated the difficulty of enforcing an assortment of uncoordinated state laws.[3] In France, the first air-related federation came into being in 1905 and promoted knowledge and safety in the domain.[4]

Although localities in the United States and elsewhere created fragmented regulations, air governance efforts at the international level began from the outset of civil aviation. In 1910, France convened the first International Air Navigation Conference in Paris, which was attended by 19 European states.[5] Nations of other continents were not invited because of the

---

[1] Garber and Herron, 2020; Ruth Stilwell, "Decentralized Space Traffic Management," paper presented at Space Traffic Management Conference 2019: Progress Through Collaboration, Daytona Beach, Fla., February 26–27, 2019.

[2] Sean Seyer, *Sovereign Skies: The Origins of American Aviation Policy*, Johns Hopkins University Press, 2021, p. 32.

[3] Seyer, 2021, p. 32.

[4] ICAO, "The Paris Convention of 1910: The Path to Internationalism," webpage, undated-l.

[5] ICAO, undated-l.

assumption that aircraft could not travel such distances. The 1910 conference concluded in disagreement over whether to treat airspace as sovereign or allow it to be open to transit.[6]

Following World War I, the victorious Allies reengaged on the international air governance debate. Most of the participants from the 1919 Paris Peace Conference signed the International Commission for Air Navigation (ICAN) agreement and ratified the accompanying Convention Relating to the Regulation of Air Navigation by the early 1920s.[7] The 1919 Paris Convention formed the foundation of air law and established the principle of sovereignty of airspace. However, the United States refused to ratify the agreement because of opposition to the League of Nations, ICAN's parent organization and a precursor to the United Nations.[8] Despite its refusal, the U.S. government came under pressure from private industry operators and insurers to adopt rules in alignment with nearby states, such as Canada.[9] Industry leaders played a major role in pushing the United States to adopt the Air Commerce Act in 1926 and a set of regulations similar to those in the ICAN agreement.[10]

The United States also engaged in air governance negotiations in its own hemisphere. U.S. delegates attended the Pan-American Convention on Commercial Aviation in Havana in 1928 along with 20 other nations from across the Americas.[11] The resulting Havana Convention created a similar but less specific set of air rules to ICAN, creating an alternative regime and fragmenting air governance.[12] The U.S. Senate delayed its ratification of the Havana Convention for several years until 1931, despite having exerted significant influence to ensure that the rules met its interests by allowing U.S. airlines to operate in the Americas.[13]

During the last years of World War II, because of the continued piecemeal nature of air governance and the proliferation of air services, various states made another push for a unified governance scheme. The United States held the 1944 Chicago Conference for the purpose of international civil aviation discussions on the establishment of world air routes and services and the creation of an international aviation council.[14] The conference was nearly derailed when its participants could not agree on the freedoms of air and frequency of opera-

---

[6] ICAO, undated-l.

[7] ICAO, undated-l; Seyer, 2021, pp. 50–54.

[8] Clement Bouvé, "The Development of International Rules of Conduct in Air Navigation," *Air Law Review*, Vol. 1, No. 1, 1930, p. 3.

[9] Seyer, 2021, pp. 126–128.

[10] Andrew Glass, "Congress Passed Air Commerce Act, May 20, 1926," *Politico*, May 20, 2013; Seyer, 2021, pp. 154–155.

[11] ICAO, "Milestones in International Civil Aviation," webpage, undated-i.

[12] ICAO, "1928: The Havana Convention," webpage, undated-a.

[13] ICAO, undated-a.

[14] David MacKenzie, *ICAO: A History of the International Civil Aviation Organization*, University of Toronto Press, 2010, pp. 30–35.

tions.[15] This issue was overcome by making available "separate agreements embodying the extent to which nations would grant each other reciprocal air rights referred to as the 'Freedom of the Air.'"[16] Thirty-six states signed the Chicago Convention in 1944, which superseded the 1919 Paris Convention and the 1928 Havana Convention and established ICAO.[17]

An International Aviation Coordinating Body

The 1944 Convention on International Civil Aviation, also known as the Chicago Convention, recognized the sovereignty of airspace above state territory and rights pertaining to international air transport.[18] Article 3 of the convention states that it is applicable to civil aircraft but not state aircraft. The parties agreed to nearly 100 articles, the first one-third of which set out standards of navigation, rights of transit, and basic air rules.[19] Most importantly, the convention created ICAO as a UN specialized agency, and the parties agreed to use ICAO to pursue as much uniformity of air rules across the globe as possible.[20] Notably, the parties did not grant ICAO any oversight authority over most commercial matters and instead placed its focus on safety, traffic management, and technical standards.[21]

ICAO's early development mirrored that of IMO. After becoming a UN specialized agency in 1947, ICAO divided labor between several plenary sessions to determine rules on air navigation, transport, finance for the organization, and legal matters.[22] The air rules these sessions produced resembled the provisions of the ICAN agreement, which facilitated quick agreement on a set of recommendations.[23] The second assembly session in 1948 finalized the structure of the organization.

---

[15] ICAO, "Freedoms of the Air," webpage, undated-d.

[16] ICAO, "Introduction," webpage, undated-h.

[17] The Chicago Conference included representatives from 52 states. Invitations were sent to 53 states, but Saudi Arabia and the USSR did not accept because of objections to the presence of other states. The USSR eventually attended but did not sign the convention until 1970. A provisional ICAO served a temporary role until sufficient ratification of the Chicago Convention in 1947. International Civil Aviation Organization, "The History of ICAO and the Chicago Convention," webpage, undated-e.

[18] ICAO, Convention on International Civil Aviation, Document 7300, December 7, 1944.

[19] ICAO, 1944.

[20] ICAO, 1944, Chapter VI, Article 37.

[21] Limiting ICAO's scope ensured support from the United States and several other nations. MacKenzie, 2010, pp. 35–52.

[22] MacKenzie, 2010, p. 74.

[23] MacKenzie, 2010, p. 69.

ICAO Structure

ICAO consists of an assembly, a council, a secretariat, and various committees (Figure 3.1). The assembly manages ICAO's budget and is "comprised of all Member States of ICAO, [and] meets not less than once in three years."[24] The council directs technical committees and includes "36 Member States elected by the Assembly for a three-year term."[25] The council's membership consists of states from three categories: "States of chief importance in air transport, States not otherwise included but which make the largest contribution to the provision of facilities for international civil air navigation, and States not otherwise included whose designation will ensure that all major geographic areas of the world are represented on the council."[26] The duties of the council also include providing annual reports to the assembly, responding to direction from the assembly, and appointing leadership for the secretariat.[27]

FIGURE 3.1

International Civil Aviation Organization Structure

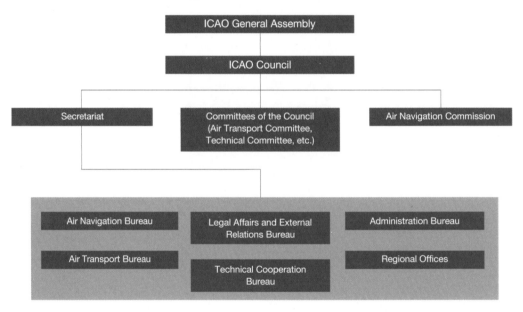

SOURCE: Adapted from Civil Aviation Authority, "ICAO's Structure and Upcoming Events in the Field of Civil Aviation (Infographics)," Government of Poland, June 11, 2013, and from discussions with Ruth Stilwell, Executive Director, Aerospace Solutions, LLC, October 3, 2022.

[24] ICAO, "Assembly," webpage, undated-c.

[25] ICAO, "The ICAO Council," webpage, undated-f.

[26] ICAO, undated-f.

[27] ICAO, undated-f.

ICAO, similar to the IMO, maintains the secretariat, a staff of experts who research policies and suggest new practices.[28] The secretariat interacts with "Invited Organizations" that include representatives from industry, from civil society groups, and from other concerned organizations to develop standards and practices.[29] Secretariat bureaus and regional offices also collect data to aid ICAO's rulemaking.[30] The ICAO apportions its expenses among its 193 members, and the assembly reviews the budget arrangements and may suspend the voting rights of any state that fails to meet its financial obligations within a reasonable time.[31]

Standards Through International Consensus

Many air rules came into being in the initial convention. Article 12 recognizes that aircraft rules and regulations are the responsibility of sovereign states but that member states should establish regulations with as much uniformity as possible to that established under the convention. For those areas over the high seas, the rules in force are those established under the convention. The articles also addressed the registration of aircraft, the display of appropriate markings, and documentation for airworthiness certificates, pilot licenses, and radio equipment.[32] Article 28 states that member states will support standard systems, such as those applicable to "communications procedure, codes, markings, signals, lighting and other operational practices."[33]

ICAO is tasked with updating technical rules only as a coordinating body that adopts global standards through the consensus of its membership body. The Air Navigation Commission plays a lead role in considering updates to the convention's annexes and new Standards and Recommended Practices (SARPs) and in submitting those proposed changes to the council for approval by a two-thirds vote.[34] Once approved, responsibility then shifts to the member states to adopt the new rules into national regulations. Aircraft operators are expected to comply with a country's regulations when operating in that country's airspace. ICAO therefore relies on the consensus of its members in agreeing to adopt any new rules that it suggests.[35] Notably, a state can file a difference when it is unable or unwilling to adopt a new standard. The council will notify the other member states, and those member states

---

[28] ICAO, "ICAO Secretariat," webpage, undated-g.

[29] International Civil Aviation Organization, "Organizations Able to Be Invited to ICAO Meetings," webpage, undated-j.

[30] ICAO, undated-g.

[31] ICAO, 1944, Chapter XII, Article 61–63.

[32] ICAO, 1944, Articles 15, 28-29.

[33] ICAO, 1944, Article 28.

[34] ICAO, 1944, Chapter X, Articles 56–57; International Civil Aviation Organization, "Making an ICAO SARP," factsheet, March 5, 2018.

[35] ICAO, 1944, Chapter IV, Articles 28(b) and 37.

may choose to limit their interactions with any state not in full compliance.[36] The ability to isolate noncompliant states in this manner acts as a soft enforcement mechanism. States are highly incentivized to adopt and abide by internationally agreed upon standards and rules to maintain both access to and interconnection with the international air domain economy.

International Accountability

The convention has several other measures to handle disputes over rules or between members. Article 84 of the Chicago Convention addresses the Council's role in settling inter-state disputes over the application of the convention and annexes.[37] Article 87 permits member states to disallow an airline to operate in their territory if the council has decided the airline is not conforming to its standards. Article 88 permits the assembly to suspend the voting power of a member state that is in default of the convention.

Outside these procedures, if a member state breaches an international agreement, then the ICAO's role is to help other member states "come up with a coordinated response"[38] and "conduct any discussions, condemnations, sanctions, etc., they may wish to pursue, consistent with the Chicago Convention and the Articles and Annexes it contains under international law."[39]

For example, the forced diversion of a Ryanair flight from its originally intended destination of Lithuania to Minsk airport in 2021 led to allegations that Belarus was not in alignment with the Chicago Convention.[40] ICAO can investigate such allegations but does not have the authority to "close or restrict a country's airspace, shut down routes, or condemn airports or airlines for poor safety performance or customer service.[41]"

Air Domain Governance

The maturity of air domain governance is reflected in the extensive and well-developed set of regulations adopted by the vast majority of states. Some air domain rules are specific to operating and piloting aircraft in a manner that might transfer to operations in space more than rules associated with the operation of vessels in the maritime domain, but both sets of rules aim to prevent collisions and therefore have relevance to the space domain. We briefly

---

[36] ICAO, 1944, Chapter VI, Article 38.

[37] Decisions of the Council can be appealed to the Permanent Court of International Justice. ICAO, 1944, Chapter IX, Article 54; ICAO, 1944, Chapter XVIII, Article 84.

[38] United Nations, "5 Things You Should Know About ICAO, the UN Aviation Agency," *UN News,* May 26, 2021.

[39] ICAO, "About ICAO," webpage, undated-b.

[40] ICAO [@ICAO], "ICAO is strongly concerned by the apparent forced landing of a Ryanair Flight and its passengers, which could be in contravention of the Chicago Convention. We look forward to more information being officially confirmed by the countries and operators concerned," Twitter post, May 23, 2021.

[41] UN, 2021.

examine some of those rules that might be applicable to space in this section. Each of these rules is documented in the ICAO Standards. For the purposes of analysis, we examine some of the U.S. Federal Aviation Regulation versions of these rules.

Technology

Technology has played a major role in enabling safer operation, and ICAO's experts develop standards for and push uniformity of new technology. Early safety concerns in the air domain focused primarily on the competency of the pilot and the airworthiness of the aircraft, which meant that safe operations depended on such rules as see-and-avoid. As congestion in the skies increased, midair collisions became a new challenge and an additional public safety concern. As World War II neared its end, the rapid development of jet engines, airframes, and radar technology created a clear need to deconflict the growing industry of international air travel.[42]

ICAO has played a role in standardizing safety measures from Air Traffic Control (ATC) services to communications and onboard sensors.[43] ICAO manages the Global Aviation Safety Plan, which provides frameworks for regional and national planning and guidance on interstate coordination.[44] The council convenes several committees and panels dedicated to building and updating standards on Communications, Surveillance, and Navigation Systems.[45]

As new technologies have developed, ICAO has become more responsive to the need for standards. One illustrative example is the 2002 Uberlingen crash, which resulted in part from insufficiently uniform regulation of onboard Traffic Alert and Collision Avoidance Systems (TCAS) and pilots' responses to TCAS instructions.[46] In response to the deadly collision, ICAO has updated TCAS version requirements and clarified ATC guidance on TCAS alerts in an attempt to standardize pilot responses to conflicting ATC and TCAS instructions.[47]

ICAO has also started working on the implementation of another key technology: Automatic Dependent Surveillance-Broadcast (ADS-B). ADS-B automatically and periodically broadcasts an aircraft's identification, GPS-determined position, and altitude to any other

---

[42] ICAO, undated-e.

[43] For an account of ATC services, see Federal Aviation Administration, "Air Traffic Control," webpage, undated. The first ATC Station was formed in 1935 by an airline consortium for aircraft separation. Of note, the first ATC services were provided in 1929, with the earliest version of those services involving one person on the ground with signal flags. Federal Aviation Administration, "First Air Traffic Controller Remembered," webpage, undated.

[44] ICAO, *Manual on the Development of Regional and National Aviation Safety Plans*, Document 10131, 2nd ed., 2022.

[45] ICAO, "Communications, Navigation, and Surveillance (CNS) Section," webpage, undated-k.

[46] ICAO, *Synopsis*, Uberlingen Accident Report, 2002.

[47] Federal Aviation Administration, *Introduction to TCAS II*, version 7.1, February 28, 2011.

aircraft with a receiver.[48] Beginning in 2004, the council convened an expert working group to assess the suitability of ADS-B for separation services, and ICAO has since begun facilitating incorporation of ADS-B into air traffic management systems.[49] The ongoing adoption of new technologies is a continuing challenge for such international organizations as ICAO and demonstrates the clear need for such organizations to maintain centralized domain expertise.

Conclusion

In the air domain, bottom-up pressure from commercial interests built as congestion and public safety concerns necessitated domestic regulations and the development of technology. As states responded to industry pressure, safety and security concerns demanded regional, and eventually international, governance mechanisms. The air domain demonstrates how bottom-up pressure can lead to a comprehensive form of international governance. ICAO's regulations have been widely adopted and provide for varying levels of need among aircraft operators, while key air domain technologies are interoperable and internationally adopted.

---

[48] ICAO, "Overview of Automatic Dependent Surveillance-Broadcast (ADS-B) Out," briefing slides, undated-l.

[49] Octavian Cioară, *Aviation System Block Upgrade (ASBU) Implementation Monitoring Report*, International Civil Aviation Organization and Eurocontrol, 2021; ICAO, *ADSB Implementation and Operations Guidance Document*, 7th ed., September 2014.

CHAPTER 4

Governance in Other Domains

Although traffic management in the air and maritime domains provides the clearest analogues for STM, valuable lessons can be learned from exploring international governance in other areas. Success stories of intergovernmental cooperation over such technical matters as telecommunications, the internet, and financial services demonstrate global governance practices that could prove applicable to STM solutions. We briefly examine the institutions that regulate and facilitate interactions within these areas and evaluate intergovernmental organizations (IGOs) writ large to draw general conclusions about which institutional features correlate with IGO effectiveness and longevity.

International Telecommunication Union

A longstanding and particularly successful example of intergovernmental coordination within a previously fragmented technical ecosystem is the International Telecommunication Union (ITU), originally founded as the International Telegraph Union in 1865 to regulate telegraph use. Assuming its current name in 1932 after its merger with the International Radio-Telegraph Union, ITU established the first international agreements on radio communications, most notably those adopted in what are now known as the ITU Radio Regulations.[1] In 1949, ITU entered the newly established UN as a specialized agency of the UN Economic and Social Council.[2] ITU covered three distinct fields: radio, telephony, and telegraphy. However, since then, the ITU has spanned broader issues in the information and communications technology (ICT) field.

As defined by ITU, "the work of ITU now covers the whole ICT sector, from digital broadcasting to the Internet, and from mobile technologies to 3D TV."[3] The organization is responsible for the global coordination of the radio spectrum and the allocation of satellite frequencies and orbital slots, for the creation of international ICT standards and cybersecurity protocols, and for facilitating the growth of telecommunication sectors in developing econo-

---

[1] ITU, "History," webpage, undated-a.

[2] ITU, "What Does ITU Do?" webpage, undated-b.

[3] ITU, undated-b.

mies (to include low- and middle-income nations). In this way, ITU fulfills the coordination role between states and satellite providers to prevent harmful interference—the signal equivalent of preventing spectrum collisions between satellites.

Currently, ITU membership spans the globe and cuts across sectors. In addition to its 193 member states, ITU has more than 900 companies, universities, research institutes, and international and regional organizations as associated or sector members.[4] ITU's funding streams—totaling 165 million U.S. dollars in budgeted revenues in 2021—are similarly diverse.[5] ITU depends on voluntary contributions for the adequate and stable funding required to execute critical spectrum and satellite coordination functions. However, ITU's central funding source—which accounts for 76 percent of total revenue—is membership fees paid by participating governments, companies, organizations, and academic institutions. This structure differs from ICAO or other UN bodies, which allow industry and NGO experts to participate only as observers that do not contribute funding or exercise any authority in governance proceedings.[6] Additional ITU revenue comes from cost-recovery activities, such as publication sales, radiocommunication filing fees, and registration fees, suggesting other nontraditional mechanisms through which a centralized STM system could be funded.

ITU is a collaborative organization relative to many UN bodies that are predominantly marked by contention and gridlock, such as the UN Security Council or the UN Human Rights Council. This cooperative nature is often an intentional byproduct of the union's democratic substructures. For instance, the Radio Regulations Board, a key rulemaking body within the ITU, is required to be composed of geographically distributed subject-matter experts who serve as stewards of public trust instead of state representatives, and members are automatically recused from ruling on orbital and radiofrequency matters involving their home states.[7] These checks and balances stem from ITU's central mission to promote, facilitate, and foster affordable and universal access to telecommunication rather than to broker geopolitical power within the ICT sector. This goal is achieved only through collaboration between state and nonstate members that willingly comply with nonbinding ITU recommendations. Interrupted cooperation would compromise the particularly interconnected ICT sector, creating natural incentives for widespread compliance with ITU standards.[8]

---

[4] ITU, *ITU-T Standardization: Committed to Connecting the World*, 2010.

[5] ITU, "How Is ITU Funded?," webpage, updated May 2022.

[6] ITU, 2022.

[7] Larsen, 2018.

[8] Jorge Ciccorossi, "Harmful Interference to Satellite Systems and the Current Challenge to GNSS," presented at Eurocontrol Stakeholder Forum on GNSS, March 4, 2021.

Several elements of the ITU's legal framework, norms, and behaviors could be applied directly to possible STM solutions, and many STM scholars have identified these analogous elements. For example, its three-level regulatory configuration has already been identified as a promising prototype for STM governance, and its multisource funding structure presents an advantageous model for financing global STM operations.[9] Because of the large and growing role that nonstate actors play in space, incorporating private representation and financial contributions into STM governance could secure buy-in from commercial actors and soften the reliance on the often-politicized government contribution model that is typically used by UN-based authorities. Furthermore, STM systems can be modeled on ITU compliance mechanisms, as well as on ITU's emphasis on collaboration and impartiality between states over competition. STM solutions, similar to ITU, should ultimately revolve around cooperation and participation among states and the private and civil sectors to ensure extensive buy-in, compliance, and institutional legitimacy and effectiveness.

Internet Corporation for Assigned Names and Numbers

A more recent entry into international technical coordination is the Internet Corporation for Assigned Names and Numbers (ICANN), a U.S.-based nonprofit founded in 1988 that has supported the development and management of the Domain Name System and IP addresses.[10] ICANN operates internationally through both internal and external legal mechanisms. However, because ICANN is headquartered in California, it must abide by both California state law and U.S. federal law, notwithstanding the fact that its functions have international effect.[11] This differentiates ICANN from ICAO, IMO, and other UN special agencies that operate in accordance with international law (e.g., treaties or customary international law rules and principles). ICANN consists of three supporting organizations and four advisory committees besides its Board of Directors.[12]

ICANN was founded in 1988 through a Memorandum of Understanding between ICANN and the U.S. Department of Commerce (DOC).[13] This relationship with the DOC gave the United States substantial influence over ICANN and therefore over eventual internet regulations. The agreement between ICANN and DOC expired in 2016, which allowed ICANN to continue evolving as an independent entity.[14] The functions of the Internet Assigned Num-

---

[9] Schrogl et al., 2018.

[10] ICANN, "What Does ICANN Do?" webpage, undated-c.

[11] Emily M. Weitzenboeck, "Hybrid Net: The Regulatory Framework of ICANN and the DNS," *International Journal of Law and Information Technology*, Vol. 22, No. 1, Spring 2014.

[12] Weitzenboeck, 2014.

[13] ICANN, "ICANN'S Early Days," webpage, undated-b.

[14] Maria Farrell, "Quietly, Symbolically, US Control of the Internet Was Just Ended," *The Guardian*, March 14, 2016.

bers Authority were assumed by ICANN in 2016, further expanding its regulatory footprint. ICANN then restructured to improve efficiencies in a global model, enhance accountability, and strengthen monitoring and enforcement mechanisms.[15] These internal regulations positioned ICANN to have a bottom-up, consensus-driven, multistakeholder approach.

ICANN works with states, international organizations, academics, technical experts, private-sector representatives, and civil society on internet naming issues.[16] ICANN strives for a more inclusive and effective multistakeholder model through communication, working groups, facilitated conversations, workshops, webinars, competitions, and more. Their expected standards of behavior and norms allow for improved operations and effectiveness.[17] The bottom-up process used by ICANN allows for broad representation of perspectives, as efforts made by other organizations and states align with ICANN's mission to regulate the internet and cyber domains.[18] ICANN is officially independent of these bodies but actively collaborates with them and is influenced and informed by their processes.[19]

Throughout ICANN's history, innovations in policies and technologies have helped the organization adapt to the ever-changing cyber and political landscape. The adoption of the multistakeholder approach has been a successful step in reinforcing ICANN's global legitimacy after the United States stepped back in 2016.[20] This legitimacy was demonstrated by China's inability to force its New IP onto this system and its subsequent need to turn to other organizations where China could have more control.[21] However, this episode also showed how states might use new technologies, such as New IP, to assert dominance and gain control in cyberspace.[22] In these efforts for increased internet autonomy, nations have exploited ICANN's Country Code Top-Level Domain (ccTLD), a program that is critical to ICANN's flexibility, to instead enforce domestic political objectives.[23] However, the ccTLDs were an

---

[15] Weitzenboeck, 2014

[16] Brian Cute, "Evolving ICANN's Multistakeholder Model: The Work Plan and Way Forward," *ICANN Blogs*, December 23, 2019.

[17] ICANN, "ICANN Expected Standards of Behavior," webpage, undated-a.

[18] ICANN, "Bylaws for Internet Corporation for Assigned Names and Numbers: A California Nonprofit Public-Benefit Corporation—ICANN," webpage, updated June 2, 2022.

[19] Weitzenboeck, 2014.

[20] L. S., "Why Is America Giving up Control of ICANN?" *The Economist*, September 30, 2016.

[21] *New IP* is generally considered to be "a set of desirable features to implement the use case described in Network 2030 (a focus group created by ITU to "carry out a broad analysis for future networks towards 2030 and beyond"). See Alain Durand, *New IP*, Office of the Chief Technology Officer, ICANN, October 27, 2020, p. 3. See also Mark Montgomery and Theo Lebryk, "China's Dystopian 'New IP' Plan Shows Need for Renewed US Commitment to Internet Governance," *Just Security*, April 13, 2021.

[22] Theo Lebryk, "The Fight over the Fate of the Internet: The Economic, Political, and Security Costs of China's Digital Standards Strategy," China Focus, April 21, 2021.

[23] Article 19, "Internet: Content Moderation at Infrastructure Level Puts Rights at Risk," blog, October 25, 2021.

important feature of the multistakeholder legitimacy-building process that helped counter-balance the perception of U.S. dominance over ICANN.[24] Compliance with the EU's General Data Protection Regulation is another example of how ICANN's cooperation with international actors has led to the development of new practices.[25]

Many lessons learned from ICANN within the internet and cyber environments can translate to the space domain and STM. ICANN's focus on bottom-up, multistakeholder approaches has cultivated widespread confidence in ICANN's regulatory practices, again highlighting the importance of active stakeholder involvement. The transition of ICANN from its perceived U.S.-dominated status to an independent—albeit still U.S.-based—multistakeholder organization in 2016 also provides an insightful analogue to the U.S.-led STM landscape and the potential path toward multilateral STM governance. Competing international interests, technologies, and capabilities have historically challenged ICANN effectiveness, but its institutional flexibility and renewed emphasis on intergovernmental consultation could be imitated by future international STM systems to help settle power disputes while preserving organizational legitimacy and effectiveness.

Society for Worldwide Interbank Financial Telecommunication

The Society for Worldwide Interbank Financial Telecommunication (SWIFT) is another international body that facilitates transactions between financial institutions worldwide. SWIFT was founded in Brussels in 1973 by several competitors of Citibank—then First National City Bank—as a countermeasure to prevent a single, private American entity from monopolizing the service of all global financial flows and to update lagging financial servicing technology.[26] SWIFT has since become the industry standard for financial communication as a result of two integral characteristics: (1) SWIFT identified a narrow yet significant

---

[24] Peter Van Roste, "ICANN71: CcTLD Governance Models—Why One Size Does Not Fit All," *Council of European National Top-Level Domain Registries*, blog, June 16, 2021l.

[25] Emmanuel Gillet, "WhoIs Data: New Response from ICANN to the European Commission," *IP Twins*, April 25, 2022.

[26] Susan Scott and Markos Zachariadis, Origins and Development of SWIFT, 1973–2009," *Business History*, Vol. 54, No. 3, June 2012. We note that for any international STM entity to be successful, it will very likely need the buy-in of most, if not all, spacefaring nations—particularly the United States, China, and Russia. Therefore, such an entity should not be in design, practice, or impression monopolized by one nation or group of nations. As described in this section, SWIFT has been viewed as Western, which could jeopardize its international reach if nations seek to create a competing mechanism. However, it is also possible that a focus by SWIFT on fundamental needs and services, such as preventing fraud and strengthening cybersecurity, can offset fears of monopolistic influence or control. This is an issue for further research because STM may be subject to the same vulnerabilities given that the United States (and U.S. companies) have an outsized role in providing SSA services.

market need and excelled in fulfilling that need, and (2) it was developed and adopted by the most-influential international institutions within the financial industry.[27]

Additionally, the organizational structure of SWIFT favorably lends itself to financial actors involved in a dynamic global environment. SWIFT provides an essential service for its users while remaining nonintrusive to transactions.[28] Although SWIFT's structure more resembles those of multinational corporations than those of ICAO or IMO, the primary oversight of SWIFT is still performed by the network of G10 central banks, granting it both governmental and nongovernmental legitimacy.[29]

However, SWIFT has faced considerable institutional and technical challenges. For instance, alternative communication methods threaten SWIFT and corresponding banking networks, potentially undermining interoperability between varied data standards and multiple closed financial systems.[30] Regulatory differences between countries further complicate payment networks.[31] Globalization and the emergence of new syntaxes and the format of messaging has made intersyntax information transfer considerably more difficult.[32]

Efforts by stakeholders across the banking and finance domain have been undertaken to develop global standards for financial communication to reduce inefficiencies in international payments. In 2004, ISO published ISO 20022 as a framework uniformly defining messaging standards across all industry operations, set for full implementation in early 2023.[33] The growing perception of ISO 20022 as a global payment standard coupled with early adopter activity in significant global economies has established ISO 20022 as a significant update and counterpart to SWIFT and a testament to SWIFT's institutional agility.[34]

The founding of SWIFT demonstrated private-sector willingness to operate collectively to develop a narrow technological service that responded to market needs. The former telex (telegraphic transfer) method of manual cross-border transactions was cumbersome and inconsistent with the technological innovations of the time, and an increasingly globalized world economy required an updated, market-driven system outside the exclusive control of

---

[27] Susan Scott and Markos Zachariadis, *The Society for Worldwide Interbank Financial Telecommunication (SWIFT): Cooperative Governance for Network Innovation, Standards, and Community*, Routledge, 2014.

[28] SWIFT, "Swift Governance," webpage, undated-a.

[29] SWIFT, "Swift Oversight," webpage, undated-b.

[30] Financial Stability Board, *Enhancing Cross-Border Payments: Stage 1 Report to the G20*, April 9, 2020.

[31] Liana Wong and Rebecca M. Nelson, "*International Financial Messaging Systems*, Congressional Research Service, R46843, July 19, 2021.

[32] SWIFT, *ISO 20022 for Dummies*, Wiley, 2020.

[33] ISO, "About ISO 20022," webpage, undated.

[34] SWIFT, 2020.

U.S. firms.[35] However, similar to ICANN, SWIFT has struggled with an enduring reputation as an intrinsically Western entity, a perception that has only intensified following its role in recent sanctions against Russia.[36] Consequently, China and Russia have pursued alternatives to SWIFT to sidestep sanctions and undermine U.S. dollar hegemony, threatening the stability of SWIFT and underscoring both the criticality and fragility of securing Chinese and Russian engagement in an STM system.[37]

Although perhaps less directly applicable than other analogues, the history of SWIFT uniquely highlights the role that private actors could play in the development of STM governance. Increasingly influential private space entities might recognize their economic incentive for improved STM and initiate international cooperation that eventually bubbles up to include participation from powerful governments. Furthermore, an STM regime involving both public and private actors could model SWIFT's internal relationships between central banks and its more than 11,000 private financial institutions. Additionally, SWIFT and ISO 20022's standardization mechanisms might offer relevant prototypes and insights for integrating SSA data and standardizing communications between satellite operators, which would be critical elements of any effective STM system.

Broader Lessons from International Governance

Trends across intergovernmental organizations (IGOs) more broadly offer additional relevant insights to STM governance. Indeed, understanding which variables tend to enhance or hinder the creation, effectiveness, and dissolution of IGOs helps assess the viability of IGO-based STM solutions and inform policy recommendations. If STM is to have a legitimate, long-standing, and effective IGO similar to ICAO, IMO, and ITU, it will need to have qualities and characteristics that support member buy-in, sustainability, and stability of operations and governance.

One telling indicator of IGO performance the opinion expressed by key stakeholders or *elites*, government officials, industry and civil society leaders, etc., about their level of confidence in an IGO's legitimacy or their perception that the body exercises its regulatory authority appropriately. Surveying these elites has generated some valuable conclusions about which features of IGOs correlate with their perceived legitimacy. One primary takeaway is that elite opinions of global institutions are heavily qualified, meaning that they express neither

---

[35] Scott and Zachariadis, 2014.

[36] Barry Eichengreen, "Sanctions, SWIFT, and China's Cross-Border Interbank Payments System," May 20, 2022.

[37] Huileng Tan, "China and Russia Are Working on Homegrown Alternatives to the SWIFT Payment System. Here's What They Would Mean for the US Dollar," *Insider*, April 28, 2022.

actively high nor critically low confidence in existing IGOs.[38] However, regional governance bodies tend to enjoy more legitimacy on average than their global or national governance counterparts, suggesting that "the middle" could be the optimal entry point for kickstarting an STM governance effort.[39] Elites also prioritize "democracy" as an institutional procedure and value "effectiveness" as the most important element of institutional performance, as opposed to such concepts as "fairness."[40]

Another significant measure of IGO success is sheer longevity. Scholars have thus examined which features are shared by the longest-standing IGOs and which qualities or dynamics commonly weaken and precipitate the demise of global governance regimes. Among the most-reliable predictors of IGO endurance are bureaucratic staff size as determined by available personnel resources, secretariat and staff salaries, the attractiveness and livability of the headquarters' localities, and other factors.[41] Furthermore, IGOs that are more technically oriented, such as ICANN or ITU, tend to outlive more norms-oriented institutions, such as those dedicated to human rights, as those institutions might be more politically sensitive and contentious topics.[42]

The primary driver of IGO dissolution is the withdrawal of key nation-states, commonly as a result of simple preference divergence.[43] For example, during World War II, the United Kingdom and Germany refused to participate in the same governance organizations. Great power departures are particularly destructive to IGOs because of their "contagiousness." Often when a major power such as the United States leaves an IGO, allies such as Canada and the United Kingdom follow.[44] Consequently, the most-successful IGOs deliberately maintain

---

[38] Jan Aart Scholte, Soetkin Verhaegen, and Jonas Tallberg, "Elite Attitudes and the Future of Global Governance," *International Affairs*, Vol. 97, No. 3, May 2021.

[39] Scholte, Verhaegen, and Tallberg, 2021, p. 877. In space, binational arrangements such as the North American Aerospace Defense Command (NORAD) prove the possibility of enhanced SSA data-sharing between allies, which could be replicated in binational or regional STM systems that snowball into larger, more-multilateral regimes as other states feel strategically or economically isolated. See James Bennett, *The NORAD Experience: Implications for International Space Surveillance Data Sharing*, Secure World Foundation, August 1, 2010, and Brian G. Chow, "How to Convince China and Russia to Join a Space Traffic Management Program for Peace and Prosperity," *SpaceNews*, January 26, 2021.

[40] Scholte, Verhaegen, and Tallberg, 2021, p. 863.

[41] Julia Gray, "Life, Death, or Zombie? The Vitality of International Organizations," *International Studies Quarterly*, Vol. 62, No. 1, March 2018.

[42] Mette Eilstrup-Sangiovanni, "Death of International Organizations. The Organizational Ecology of Intergovernmental Organizations, 1815–2015," *Review of International Organizations*, Vol. 15, No. 2, April 2020.

[43] Eilstrup-Sangiovanni, 2020.

[44] Inken von Borzyskowski and Felicity Vabulas, "Hello, Goodbye: When Do States Withdraw from International Organizations?" *Review of International Organizations*, Vol. 14, No. 2, June 2019.

lock-ins for continued U.S., Chinese, and Russian involvement to prevent these often-fatal exoduses. STM would very likely need to make the same effort. Experts agree that improved space traffic governance will necessitate strong involvement from the three space powers.[45] Since the drafting and acceptance of the four primary space treaties, success stories of global governance in space have been few and far between, with the United States, China, and Russia seldom seeing eye to eye when it comes to space.[46] Several factors are at play: advancements in military space technology that increase the security implications of space, the rise of additional space powers (specifically China, which offers Russia a non-Western space partner and counterweight to U.S. space hegemony), an expanded presence of private space actors, a general erosion of confidence in international institutions, preference divergence, and deterioration of great power relations, among others.[47]

Consequently, the challenge of striking cooperation between great space powers over an STM regime would be difficult. However, other cross-domain analogues might offer some optimism and potential ways forward in securing engagement from the United States, China, and Russia despite the heavily federated STM state-of-play. One possible corollary is arms control, an area similar to space in that it involved technological competition that was initially exclusive to the United States and USSR and eventually regulated by a groundswell of UN activity but that has since largely disintegrated because of shifting geostrategic realities. Indeed, from 1960 through the end of the century, sustained political investment in arms control as a mechanism for preventing the unthinkable resulted in a robust patchwork of international commitments (Strategic Arms Limitation Talks [SALT I, SALT II], Strategic Arms Reduction Treaty [START I], etc.). However, as in space, attitudes toward multilateral governance have soured in the decades since, with a significant share of the arms control regime eroding without replacement. The notable exception to this shift has been the New START Agreement, which was signed in 2010 and extended for five years in 2021. Although nuclear arms control and space are far from equivalent diplomatic arenas, the anomalous perseverance of New START—although it is an imperfect agreement—against strong headwinds demonstrates the possibility, however narrow, of agreement among great powers to new global governance within the similarly mired space domain.

---

[45] Mir Sadat and Julie Siegel, *Space Traffic Management: Time for Action*, Atlantic Council, August 2022.

[46] The four primary space treaties are Treaty on Principles Governing the Activities of States in the Exploration and Use of Outer Space, Including the Moon and Other Celestial Bodies, January 27, 1967; Agreement on the Rescue of Astronauts, the Return of Astronauts and the Return of Objects Launched into Outer Space, April 22, 1968; Convention on International Liability for Damage Caused by Space Objects, March 29, 1972; and Convention on Registration of Objects Launched into Outer Space, January 14, 1975.

[47] Matthew Looper, "International Space Law: How Russia and the U.S. Are at Odds in the Final Frontier," *South Carolina Journal of International Law and Business*, Vol. 18, No. 2, 2022.

Although buy-in to IGOs from China, Russia, and Iran has proven especially difficult to secure, the United States in fact leads the world in IGO withdrawals.[48] The United States has historically declined or withdrawn from global commitments for a myriad of reasons, most often as a result of domestic political pressure. In the case of arms control, this has manifested as U.S. politicians simply reorient their security priorities (as they did in the wake of September 11, 2001) or more directly question the value of arms control agreements in the face of Russian violations or general aggression. Regardless of motivation, many consider the U.S. Senate's 1999 failed ratification of the Comprehensive Nuclear-Test-Ban Treaty to have precipitated the agreement's eventual demise, representing the first major arms control domino to fall. The U.S. withdrawal from the Anti-Ballistic Missile Treaty in 2002 accelerated the trend, with Russia following suit in 2007 by rolling back its participation in the Conventional Forces in Europe Agreement and ending the Cooperative Threat Reduction program in 2012. The collapse was completed by the Trump administration withdrawing from the UN Arms Trade Treaty, the Intermediate-Range Nuclear Forces Agreement, and the Open Skies Treaty.[49]

In space, U.S. political forces have similarly hindered multilateral cooperation, as in the case of the Wolf Amendment, which has forbidden NASA from direct collaboration with China's National Space Administration since 2011 and further cements space as a primarily competitive domain for great powers, even for civilian organizations.[50] Of course, the United States hardly shoulders sole responsibility for the current geopolitical space environment (or for the disintegration of the arms control infrastructure). Russia and China routinely antagonize the United States and Western allies in space and regularly reject and undermine U.S. and UN efforts to facilitate space multilateralism.[51] Thus, STM efforts must specifically acknowledge the unique challenge of motivating U.S. participation, in addition to Chinese and Russian participation.

Of course, IGOs can and have survived and experienced success without featuring every quality that is historically correlated with legitimacy and longevity. However, an IGO-based STM approach would enjoy the highest likelihood of sustained success by incorporating key strategic lessons: suturing democratically structured regional bodies into a UN-based organization, prioritizing strong technical cooperation, ensuring high bureaucratic staff size with-

---

[48] Jon C. W. Pevehouse, Timothy Nordstrom, Roseanne W. McManus, and Anne Spencer Jamison, "Tracking Organizations in the World: The Correlates of War IGO Version 3.0 Datasets," *Journal of Peace Research*, Vol. 57, No. 3, June 2020.

[49] See Steven Miller, *Hard Times for Arms Control: What Can Be Done?* The Hague Center for Strategic Studies, February 2022.

[50] See Bruce W. MacDonald, Carla P. Freeman, and Alison McFarland, *China and Strategic Instability in Space: Pathways to Peace in an Era of US-China Strategic Competition*, United States Institute of Peace, February 2023.

[51] See Theresa Hitchens, "At UN Meeting, Space Cooperation Picks up Momentum, but Moscow and Beijing Play Spoilers," *Breaking Defense*, February 3, 2023.

attractive pay and locality, identifying Russian and Chinese participation as mission-critical, and sustaining Russian, Chinese, and U.S. involvement with deliberate lock-in mechanisms and incentives for all three space powers.[52]

---

[52] Keeping the United States, China, and Russia at the table will likely require clearly aligned economic incentives (i.e., gaining access to foreign space markets requires participation in an STM regime), significant representation from each country in a possible IGO's secretariat and cadre of resident experts, and a council-type structure (e.g., ICAO Council or UN Security Council). These mechanisms would ensure organizational privileges and influence (to the extent practicable and equitable) for great space powers given their significant proportional interests and investments in space beyond other nations. See Gilles Doucet, "Characteristics of Governance for Beyond National Jurisdiction: Lessons for Future Outer Space Governance," *Astropolitics*, Vol. 20, No. 2–3, May–December 2022, and Brian Chow, "Space Traffic Management in the New Space Age," *Strategic Studies Quarterly*, Vol. 14, No. 4, Winter 2020. Additionally, as part of the agreement or treaty creating such an organization, disincentives could be included for withdrawal—i.e., affirmatively reducing lack of access to SSA and STM data and information, reduced or removal of voting rights, or reduced opportunity for SME participation and input.

CHAPTER 5

Insights and Recommendations

We draw insights and recommendations for next steps in the implementation of an international STM system based on the previous reviews of the evolution of governance structures in other domains and areas, the work done to consider options for STM, and lessons from the history of other IGOs, To manage such a system, we find that the most promising option is for the international community to create an ISTMO.

Key Insights

The World Is Approaching a Space Traffic Management Tipping Point

First, it appears that the world is approaching a tipping point for STM. Compared with the maritime domain, there has been much less time to develop the underlying norms and motivations for an international system. However, closer reflection indicates that the major changes in both the maritime and air domains took place in less than 50 years. These changes occurred after centuries of evolution of maritime technology, treaties, and norms, and less than 50 years of evolution of air domain equivalents. Figure 5.1 summarizes these overall changes for the two traditional domains and the space domain. Table 5.1 summarizes key governance aspects of each domain, with a focus on traffic management.

Chronologically, one could argue that space is already overdue for the creation of an international governance system similar to IMO and ICAO. More importantly, the rapid growth in the space domain, especially since the early 2000s, is reflective of the accelerating trends in technology and domain usage that catalyzed the development of similar governance systems in the air and maritime domains.[1] Besides the growing importance of space-to-human activity, there is also the growing urgency associated with the increase in the number of objects (both maneuverable and nonmaneuverable debris) in orbit.

There remains a spectrum of views internationally regarding the feasibility and advisability of an actual ISTMO. During our workshops, European and Asian Pacific experts agreed that the ISTMO concept was a potentially feasible solution more readily than U.S. attendees

---

[1] Bruce McClintock, Katie Feistel, Douglas C. Ligor, and Kathryn O'Connor, *Responsible Space Behavior for the New Space Era: Preserving the Province of Humanity*, RAND Corporation, PE-A887-2, 2021.

FIGURE 5.1

Maritime, Air, and Space Governance Development Timelines

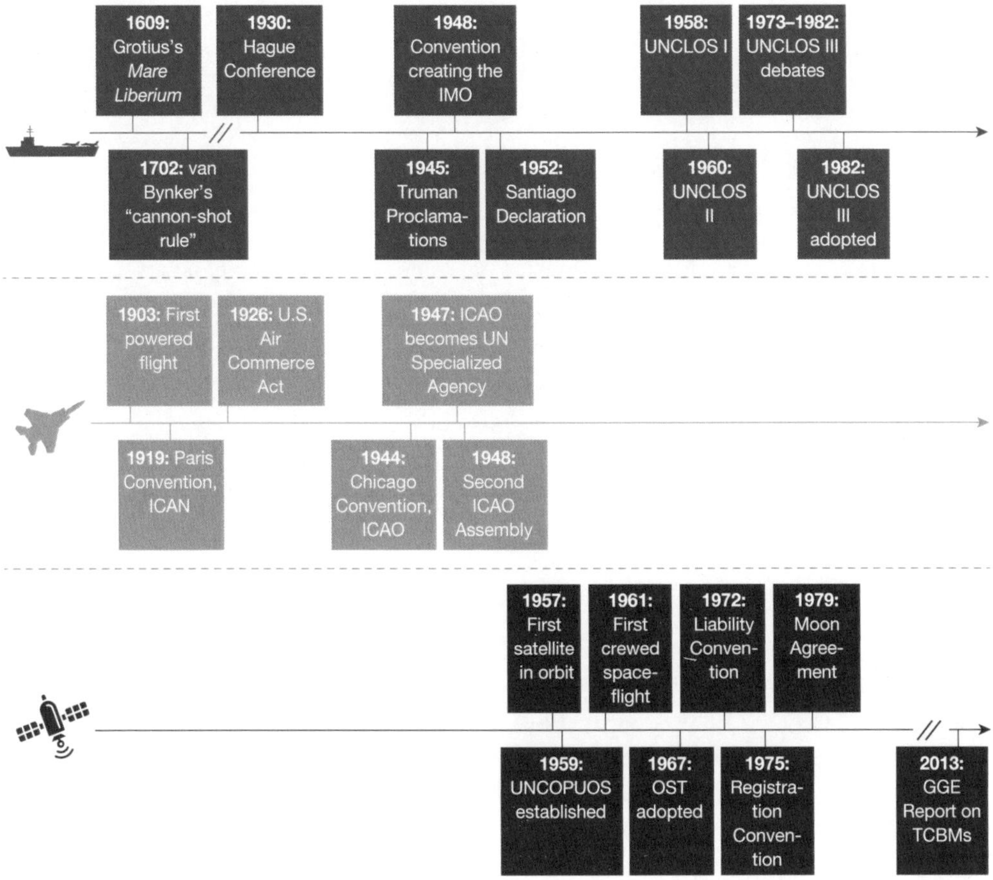

SOURCE: McCormick, Ligor, and McClintock, 2023, p. 10.

did.[2] All European workshop participants further agreed that ISTMO is an advisable solution, while the majority of Asia Pacific and U.S. workshop participants did not.[3] Although this small sample size is not conclusive, it might reflect some shift toward more support for an ISTMO than was present in the past. In any case, history in other domains and areas indicates that an international organization is likely to eventually emerge out of necessity and because the negative consequences of waiting (e.g., a major collision or incidence of conflict in space resulting from an STM failure) would be significant.

---

[2] For the workshop, we defined *feasible solution* as a solution that would address the needs of the STM system in a manner that is technologically and functionally implementable.

[3] For the workshop, we defined *advisable solution* as a solution that would address the needs of the STM system in a sensible way from a geopolitical and socioeconomic perspective.

TABLE 5.1

Key Aspects of Maritime, Air, and Potential Space Governance and Traffic Management

	Maritime Domain	Air Domain	Space Domain
Key IGOs	• UNCLOS • IMO • ITMOS • ICJ	• ICAO	• STM might emerge as UN-based entity or amalgamation of regional entities
Historical challenges	• Waters that are disputed between states • No standardized claims to water similar to those that exist for land • No standardized dispute-resolution mechanisms worldwide	• Lack of standardized air traffic metrics • Lack of formal agreement to permit interstate air travel	• No formal body or mechanism to standardize space traffic procedures, prevent collisions, or resolve disputes
Collision environment	• Vessels vary in size (from couch-sized to the size of five football fields) with slower speeds (1 to 60 knots), making avoidance easier • Debris usually sinks or drifts with currents • High-level human involvement	• Vessels are generally smaller, but speed is far greater than in maritime domain (75 to 500 knots) • No lingering debris created that increases congestion • High-level human involvement	• Vessels vary in size (from toaster-sized to school bus–sized) and move at very high speeds (6,000 to 15,200 knots) • Debris avoidance affects fuel margins, shortening mission life
Limitations	• No strict enforcement or delineation on high seas compared with near the shoreline • Regulations for automated vessels still nascent	• No central investigatory body for collisions in international airspace • ICAO involvement only as requested by state leading investigation	• All of space faces limitations equivalent to high seas or flight outside sovereign state airspace
Solutions	• UNCLOS delineates clearly how sovereign waters are divided among states • ICJ, IMO, and ITMOS resolve disputes and enforce standards	• ICAO agreement involves large buy-in to delineate common metrics and formal agreement to allow for interstate air travel	• Potential for centralized agreement with a large buy-in • Dispute resolution can be integrated with ICJ or similar adjudicative body
Governance structure	• Assembly of all states elects council that manages bureaucratic committees • Council structure represents diversity of regions and domain interests	• Assembly of all states elects council that manages bureaucratic committees • Council structure represents a diversity of regions and domain interests	• Will likely require buy-in from even small states with some broad-based voting and deliberative system that includes United States, EU, China, and Russia

There Is Already Extensive Research on Space Traffic Management Process and Governance Options

The research into the need for STM organizations, process to implement STM organizations, and organizational and funding options for STM organizations is already extensive. As noted in Chapter 1 and shown in Table 1.1, there have been more than a dozen major conferences, reports, and papers published going back 40 years and discussing the necessity of and options for an STM system. A widely accepted definition of STM dates from 2006, and several reports provide extensive detail on options for STM system implementation and organization. More recently, every session of the UNCOPUOS LSC since 2016 has included formal deliberations on STM. The 2022 report is the latest to provide extensive detailed analysis and recommendations for an international space traffic management system.[4] It is time to act on past research and the governance groundwork already put in place by UNCOPUOS.

A Future Space Traffic Management Governance System Needs Legitimacy to Endure and Be Effective

Analysis of other domains and areas and analysis of IGOs in general demonstrate that an organization needs legitimacy to survive. *Legitimacy* in the context of international governance does not have an agreed-upon definition. However, most definitions of *legitimacy* have at least similar elements: a collective belief in the same social construct, acceptance and justification of shared rule by a community, the justification of actions by those affected according to reasons they accept, governance structures established in accordance with accepted rules and principles, and a voluntary acceptance of a political sovereignty based on the belief that doing so is rightful, just, and in accordance with prevailing social values.[5] We define *legitimacy* in the context of an ISTMO as the belief in the rightful use of authority by an institution, and it is operationalized as the observable behavior of either deference to the institution or opposition to it.[6]

Thus, for an ISTMO to be successful, it must be chartered with a set of authorities to which the international community, particularly spacefaring nations, will defer. But those authorities cannot be so broad or ill-defined that they create persistent opposition or lack of compliance. To effectively influence STM, an ISTMO will need to be empowered with authorities related to (1) SSA and STM measurement, (2) the institutionalization of STM expertise, (3) rules development and establishment via an accepted voting regime (some selec-

---

[4] UNCOPUOS, 2016.

[5] See, generally, Steven Bernstein, "Legitimacy in Intergovernmental and Non-State Global Governance," *Review of International Political Economy*, Vol. 18, 2011, pp. 17–51; Jens Steffek, "The Power of Rational Discourse and the Legitimacy of International Governance," 2000.

[6] See definition of *legitimacy* in Ian Hurd, "Legitimacy and Contestation in Global Governance: Revisiting the Folk Theory of International Institutions," *Review of International Organizations*, Vol. 14, December 5, 2018, p. 718.

tion or combination of selection among consensus, majority, and weighted voting), (4) compliance and enforcement mechanisms, (5) building a strong bureaucratic organization that can deliver quality outputs, and (6) conflict adjudication mechanisms.[7] Member states must accept the ISTMO's exercise of authority in these six areas as legitimate to ensure endurance and efficacy.

In addition to these six areas, the evidence indicates the importance of strong technical cooperation and collaboration between ISTMO members. Technology in traditional domains has proven to be a force multiplier, and the same potential exists for space if actors and stakeholders are willing to share the information necessary to make STM workable. Technical cooperation and collaboration are important because, as shown in the cases of IMO and ICAO, lasting IGO systems need data, information, and measurements that are reliable and trustworthy to ensure situational awareness, inform decisionmaking, and help resolve disputes. These characteristics can be achieved only if members of the IGO work together on technical aspects. Moreover, successful IGOs have established and maintained domain expertise that informs rulemaking through sustained technical cooperation and collaboration. An ISTMO's legitimacy will also be a function of its ability to facilitate technical cooperation and collaboration and apply the outputs to STM.

Finally, and perhaps most critically, ISTMO legitimacy will necessitate buy-in from key space powers, including the United States, China, and Russia, while including extant regional bodies, nonspacefaring nations, and low- and middle-income countries, and eventually connecting them via an international UN-based organization. The air and maritime domains offer lessons on how the international community can develop an international organization to address STM. In both domains, international norms and customs preceded hard codifying of actual regulations, a process that is already underway in the space domain.[8] An ISTMO with strong legitimacy must be able to convert current norms of behavior and activities that support STM into a coherent body of rules and regulations by which space actors will abide.

Existing Activities Show That a Bottom-Up Approach Is Already Underway for STM

The recent emphasis of STM in various regions demonstrates that there might already be some movement toward broader STM governance structures. Research indicates that bottom-up or minilateral approaches tend to start necessary debates over rule specifics, which is essen-

---

[7] In addition to the broader lessons for IGO strength mentioned in Chapter 4, including the fact that larger bureaucratic IGOs tend to be more enduring and effective, research indicates that the bureaucracy must be sufficiently robust to function in a quality manner. Member satisfaction with the functioning of an ISTMO bureaucracy will be essential to its success. See Francis Fukuyama, "Governance: What Do We Know, and How Do We Know It?" *Annual Review of Political Science*, Vol. 18, December 21, 2015, p. 92.

[8] The ongoing development and codification of cybernorms offers an additional, more modern corroboration of this dynamic. See Martha Finnemore and Duncan B. Hollis, "Constructing Norms for Global Cybersecurity," *American Journal of International Law*, Vol. 110, No. 3, July 2016.

tial for regulatory development, agreement, and long-term compliance.[9] Such efforts as the EU Approach to STM demonstrate that intergovernmental efforts are already being made in this area.[10] These efforts and research that indicates that regional governance bodies tend to enjoy more legitimacy on average than their global or national governance counterparts suggest that leveraging the existing approach could be the optimal entry point for kickstarting an STM governance effort.[11] It is also important to note that bottom-up approaches require state-led governance efforts, such as the creation of bilateral or multilateral agreements and organizations. Research regarding the commercial spaceflight industry indicates that commercial or industry-led standard development efforts alone are unlikely to produce agreement and implementation of regulatory standards.[12]

A Viable International System Requires Adequate Expertise and Funding

Examples of other domain management organizations have adequate member state participation and appropriate staffs and resources for the tasks assigned to them. ICAO's budget for 2020 to 2022 was 322 million Canadian Dollars (CAD),[13] with a total staff of 908 individuals in 2021.[14] IMO's budget for 2021 was 44.29 million British Pounds (GBP), with a total staff of 320. [15] The necessary budget and staffing level to effectively operate an ISTMO would need to be determined through further study. IMO and ICAO, which are primarily funded by member nations, could be starting points for this research. Another starting point option is

---

[9] We define *minilateral* or *minilateralism* as "an informal or formal grouping of three to five states that aim to coordinate their strategic agendas and facilitate functional cooperation in particular issue areas." See Kei Koga, "A New Strategic Minilateralism in the Indo-Pacific," *Asia Policy*, Vol 17, No. 4, October 2022, p. 28; Ian A. Christensen and Christopher D. Johnson, "Putting the White House Executive Order on Space Resources in an International Context," *Space Review*, April 27, 2020. See also, in the climate regulation context, Rafael Leal-Arcas, "Top-Down and Bottom-Up Approaches in Climate Change and International Trade," paper presented at the 10th Annual Conference of the Euro-Latin Study Network on Integration and Trade (ELSINT), October 19–20, 2012, pp. 1–4; Yu Hongyuan and Yu Bowen, "Global Climate Governance: New Trends and China's Policy Options," *China International Studies*, Vol. 61, December 12, 2016.

[10] EUSTM, "STM Is Getting a Prominent Place in European Space Policymaking," June 29, 2022.

[11] See, generally, Robert Faulkner, "A Minilateral Solution for Global Climate Change? On Bargaining Efficiency, Club Benefits, and International Legitimacy," *American Political Sciences Association*, Vol. 14, No. 1, March 2016.

[12] See Douglas C. Ligor, Benjamin Miller, Maria McCollester, Brian Phillips, Geoffrey Kirkwood, Josh Becker, Gwen Mazzotta, Bruce McClintock, and Barbara Bicksler, *Assessing the Readiness for Human Commercial Spaceflight Regulations: Charting a Trajectory from Revolutionary to Routine Travel*, RAND Corporation, RR-A2466-1, 2023.

[13] ICAO, "Budget of the Organization for 2020-2021-2022," webpage, 2019.

[14] ICAO, "Progress on ICAO Accountability and Transparency," webpage, 2021.

[15] IMO, *Financial Report and Audited Financial Statements for the Year Ended 31 December 2021*, December 31, 2021, p. 3; IMO, 2021, p. 17.

the ITU, which has a total budget of $165 million USD and relies on fees paid by participating companies, nongovernment organizations, and academic institutions, in addition to member governments. As state buy-in grows, so will the necessary level and amount of expertise to run any new organization and to develop technical rules that allow the eventual adjudication of conflicts and establishment of compliance mechanisms. Other domain organizations, such as IMO and ICAO, have historically used funding provided by states, but there are other models available, as discussed in the Recommendations section.

Recommendations

Key Space Powers Should Formally Start the Discussion to Establish an International Space Traffic Management Organization

The time has passed to just study the problem of STM, and it is now time to convene the appropriate organizations to decide on the path forward. The goal should be an international STM convention within the next five years that sets specific milestones for implementation within the next ten years. The time is now to actively develop STM rather than waiting for loss of key orbital shells because of collision-generated debris that limits use of valuable orbits for decades or longer. Past efforts have called for a Chicago Convention for Space to no avail, and UN Discussions on STM at the UNCOPUOS LSC since 2016 have been helpful but have not generated the necessary momentum for more-significant progress.[16]

Key space players could use the example of the UN Resolution 75/36 process to kick-start the international discussion and move from lower-level discussions to international discussions that avoid claims of lack of coordination or regional bias.[17] Early discussions should include not only major space powers but also key regional stakeholders, nonspacefaring nation representatives, industry, academia, and NGOs. Regional circles of trust (i.e., groupings of allies or like-minded nations) can be the key input node for developing an ISTMO

---

[16] Both UNOOSA and UNCOPUOS are hindered by their functional and structural characteristics with respect to making progress on a single, discrete issue such as STM. Their jurisdictions cover all areas of the space domain—exploration, technology, sustainability, economic and social development, governance, etc. Additionally, they are consensus-based organizations. Thus, outcomes are the result of complex trade-offs and deliberations that are difficult to achieve among a large and diverse group of member nations. Because of this combination of characteristics, solving single and distinct problem such as STM becomes extremely difficult. See, generally, Andrew Guzman, "The Consent Problem in International Law," working paper, University of California Berkley Program in Law and Economics, March 10, 2011.

[17] UN, "Reducing Space Threats Through Norms, Rules, and Principles of Responsible Behaviours," resolution, A/RES/75/36, December 7, 2020. This resolution was a call to member states to develop further rules for space. This resolution was instrumental in the creation of the UN Open-Ended Working Group for the purpose of developing solutions to reduce space threats. See UN, "Reducing Space Threats Through Norms, Rules, and Principles of Responsible Behaviours," A/RES/76/231, adopted December 24, 2021. The Open-Ended Working Group was organized into three sessions in May 2022, September 2022, and August 2023. See UN "Open-Ended Working Group on Reducing Space Threats," webpage, 2022.

while maintaining participation and buy-in from nations that might not participate auton-
omously because of lack of resources or fear of marginalization from other multinational
blocs. This start-in-the-middle approach could be employed to build SSA and STM capacity at the regional level first before feeding into the ISTMO. The role of industry and NGOs
is important given the extensive contributions of both to the development of SSA and STM
technology, frameworks, and proposals.[18] This approach would gain the full benefit of representation by diverse stakeholders, nations, and regions while mitigating potential capture of
the ISTMO by elite, powerful, and well-resourced spacefaring nations and entities.

The Space Traffic Management Convention Should Learn from Past Successes

As we have examined, several organizational options have been analyzed in previous literature, and there is no one perfect model for an ISTMO. However, there are some clear best-
practice considerations that should be incorporated. First, the organizational design should
be cooperative, collaborative, and inclusive, and its creation and design should be based on
consensus to ensure legitimacy. IMO and ICAO have participation from most states across
the globe. Each has an assembly of all members and a council that draws membership from
three categories to ensure diversity of representation. In both cases, the structures of the
councils incorporate both geographic and financial interest diversity while having processes
and procedures to bring about convergence and agreement on rules.

Second, the actual creation of rules and enforcement should be driven by a less restrictive
process to ensure representation, but one that also incorporates voting rules (e.g., majority or
weighted rules) to prevent gridlock. These voting rules will allow enforcement equality and
ensure incorporation of diverse views. Both IMO and ICAO allow bottom-up development of
rules and participation by nonstate actors, including private companies. Similarly, an ISTMO
should have structures and mechanisms built into its committee, subcommittee, and working group processes to ensure that nonstate actors are effectively integrated. This integration
is a critical element because nonstate actors, particularly in industry, are primary operators in
space and will be able to provide the most-accurate data, information, analysis, and insights
for governance decisionmaking. Civil society groups and academic experts similarly have an
important perspective to offer. Moreover, these structures and mechanisms should allow not
only for nonstate actor inputs, but also for industry feedback on current and proposed regulations or other aspects of governance to ensure relevance and encourage buy-in and long-term
acceptance. Although the STM convention should be responsible for the STM organizational
design, it should model that design around these characteristics to build on past successes
and garner buy-in from spacefaring and other nations.

---

[18] See Global Industry Analysts, *Global Space Situational Awareness (SSA) Industry*, January 2023; Lal et al.,
2018.

The Global Space Community Needs to Gather and Grow the Cadre of Experts to Staff an International Space Traffic Management Organization

The demonstrated value of technology as a tool to make interactions in a domain safer indicates that the ISTMO community will need to gather and grow a cadre of technical space experts. As discussed, institutionalized expertise is a key factor in not only technology uptake but also the legitimacy, efficacy, and longevity of IGOs.[19] Without sufficient expertise, ISTMO members are likely to ignore or be apathetic to ISTMO processes, decisions, and rulemaking. To create a centralized pool of expertise, any new international organization will need to have adequate staff size and to be appropriately resourced to compensate the experts. The STM convention should consider the organizational construct, size, and technical expertise needed for any future ISTMO.

Future Research Should Consider Alternative Funding Mechanisms for an International Space Traffic Management Organization

Considering the challenges other international organizations have faced with state funding, there should be more research dedicated to identify alternative funding mechanisms that could be considered for an ISTMO.[20] As discussed above, IMO, ICAO, and ITU have methods to apportion funding between their member states, and IMO specifically accounts for overall use of or access to the domain in determining contributions.[21] These would be traditional models of funding for an ISTMO.

Other nontraditional funding mechanisms could also be explored. Some examples that could have merit for the future include concepts such as orbital-use fees (OUFs) or Tradeable Satellite Performance Bonds (TSPBs). OUFs are effectively a tax on satellites and could not only increase the value of the space industry but could also provide a user-based funding mechanism for a future ISTMO.[22] Similarly, TSPBs are a market-based instrument for limiting the growth of space debris and incentivizing more-sustainable and more-efficient uses of orbital space.[23] Like OUFs, TSPBs price the sustainability of orbital-use behaviors;

---

[19] For an example from the energy domain, see Johanne Grøndahl Glavind, "Bureaucratic Power at Play? The Performance of the EU in the International Atomic Energy Agency," *European Security*, Vol. 24, No. 1, 2015.

[20] For an example involving the UN's World Health Organization, see Nilima Gulrajani, Sebastian Haug, and Silke Weinlich, "Fixing UN Financing: A Pandora's Box the World Health Organization Should Open," ODI, January 26, 2022.

[21] See, generally, IMO, 2021, p. 3; IMO, 2021, p. 17.

[22] Akhil Rao, Matthew G. Burgess, and Daniel Kaffine, "Orbital-Use Fees Could More than Quadruple the Value of the Space Industry," *Proceedings of the National Academy of Sciences*, Vol. 117, No. 23, June 9, 2020.

[23] Our conception of TSPBs is based on, and adapted from, the following research: Zachary Grzelka and Jeffrey Wagner, "Managing Satellite Debris in Low-Earth Orbit: Incentivizing Ex Ante Satellite Quality

more-sustainable behaviors incur lower costs. The net cost of the TSPB to the operator who posts the bond is only the sum of damage charges accrued over the satellite's lifetime and any financing costs. The rest of the deposit is either returned to the operator over time or collected immediately (net of expected damage charges) through market trading. Counterparties who purchase the bond will not pay more than what they expect to receive from the bond over its lifetime. Unlike OUFs, TSPBs create a tradable asset that produces a stream of cashflow for the holder and can be used to smooth business risk. TSPBs might incentivize both better behaviors from operators and innovations in and demand for such services as SSA and active debris removal.

Notwithstanding the opportunity to generate revenue, these nontraditional mechanisms would not likely be a direct tool for an ISTMO. OUFs and TSPBs act as a tax or fee directly imposed to an entity conducting space operations. There is no precedent for an international governmental organization to directly tax or impose such fees on individuals or individual entities under the sovereign jurisdiction of a state.[24] Because states would be unlikely to agree to such a system, these mechanisms would most likely need to be employed at the national level. Revenues from these mechanisms could then be used to pay ISTMO member fees.

Moreover, governments that participate in an ISTMO can monetize their legacy space debris stocks by posting the bonds themselves, then allocating them to entities they believe can dispose of the objects. Again, this would be a state revenue-generating mechanism. States could then use this revenue to pay membership fees and costs into the ISTMO. These bond allocations could raise revenues through bond auctions or through other revenue-generating methods consistent with the government's objectives. Regardless, monetizing the legacy space debris stock immediately raises the value of developing active debris removal and in situ assembly and manufacturing technologies through the potential of receiving the bond deposit amount.

---

and Ex Post Take-Back Programs," *Environmental and Resource Economics*, Vol. 74, 2019; Derek Lemoine, "Incentivizing Negative Emissions Through Carbon Shares," working paper, National Bureau of Economic Research, November 2021; and Organisation for Economic Co-operation and Development, *What Have We Learned About Extended Producer Responsibility in the Past Decade? Case Study—Chile*, 2014.

[24] The International Seabed Authority, an IGO created by UNCLOS, has a mechanism for collecting royalties and payments from contracts that it issues for deep seabed mining. See International Seabed Authority, "Equitable Sharing of Financial and Other Economic Benefits from Deep-Sea Mining," January 2022. However, the International Seabed Authority has yet to issue such contracts or collect such royalties and payments. This is in contrast to space, where satellite operators have been using space for decades and paying taxes and fees to their respective states, when applicable.

Workshop Findings

Workshops

In November 2022, we held three 90-minute virtual workshops across three distinct regions—Europe, Asia Pacific, and the United States—to discuss our research approach and proposed STM solution. To ensure a diverse set of participants across government, academia, and nongovernment organizations, we used a mix of purposive and snowball sampling. Workshop attendance ranged from seven to nine participants, excluding RAND team members (Table A.1). The smaller group size allowed for a robust group discussion. Workshops were held under Chatham House Rules and not recorded. A RAND team member captured participant comments for note-taking purposes.

During the three workshops, a RAND moderator briefed the ISTMO concept and solicited feedback from participants. Utilizing the software tool SLIDO, participants rated ISTMO on two dimensions: feasibility and advisability. *Feasibility* was described as the ability of the international community to implement the solution from a technology and functionality perspective, and *advisability* was described as the degree to which the solution would address the needs of the system in a sensible way from a geopolitical and socioeconomic perspective. Participants rated the statements "ISTMO is a feasible solution" and "ISTMO is an advisable solution" on a scale from 1 to 5, on which 1 = strongly disagree, 2 = disagree, 3 = neither disagree nor agree, 4 = agree, and 5 = strongly agree). Responses were grouped by their ratings as negative (disagree and strongly disagree), neutral (neither disagree nor agree), or positive (agree and strongly agree).

As shown in Figure A.1, more European and Asian Pacific workshop attendees agreed that the ISTMO concept was a potentially feasible solution than did U.S. workshop attendees, who cited technical issues with the coordination and data collection for such an effort that would

TABLE A.1
Workshop Details

Workshop	Date	Number of Participants
Europe	November 7, 2022	8
Asia Pacific	November 8, 2022	7
U.S. Plenary	November 15, 2022	9

FIGURE A.1
Workshop Feasibility Rating

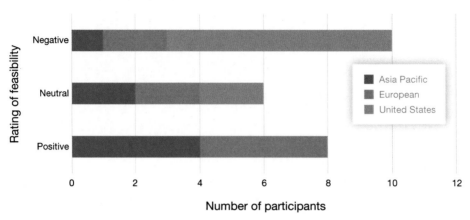

Number of participants

require "something with almost God-like thought." Asia Pacific workshop participants discussed the difference between an organization chartered with coordination, which would be more similar to ICAO, and an organization that has an operational charter, which was viewed by some as more difficult to implement. Some European workshop participants indicated that the concept was feasible, but they noted that military operations that seek exceptions to such a system could be a barrier.

Although all European workshop participants agreed that an ISTMO is an advisable solution, the majority of Asia Pacific and U.S. workshop participants did not agree that it was advisable (Figure A.2). European workshop participants' views that ISTMO is a poten-

FIGURE A.2
Workshop Advisability Rating

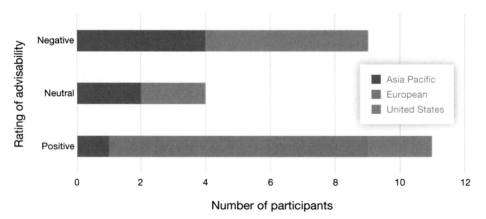

Number of participants

tially advisable solution align with the EU's current approach to start with a regional STM effort.[1]

U.S. and Asia Pacific workshop attendees discussed geopolitical challenges, such as the challenge that "[w]e would struggle to even get everyone to sit at the same table to get the conversation started." Another U.S. workshop attendee noted that, "With the asymmetry of U.S. use in space, we need to focus on ourselves before we look internationally."

---

[1] EU, "An EU Approach for Space Traffic Management," fact sheet, February 15, 2022.

Abbreviations

AIAA	American Institute of Aeronautics and Astronautics
AIS	automatic identification system
ASAT	antisatellite
ATC	Air Traffic Control
COLREGs	Convention on the International Regulations for Preventing Collisions at Sea
CONFERS	Consortium for Execution of Rendezvous and Servicing Operations
ESA	European Space Agency
EU	European Union
GGE	Group of Governmental Experts
IAA	International Academy of Astronautics
ICAN	International Commission for Air Navigation
ICANN	Internet Corporation for Assigned Names and Numbers
ICAO	International Civil Aviation Organization
ICT	information and communications technology
IGO	intergovernmental organization
IMO	International Maritime Organization
ISO	International Organization for Standardization
ISTMO	international space traffic management organization
ITU	International Telecommunication Union
LEO	low earth orbit
LSC	Legal Subcommittee
LTS	Long-Term Sustainability
NASA	National Aeronautics and Space Administration
OST	Outer Space Treaty
OUF	orbital-use fee
RPO	rendezvous and proximity operation
SOLAS	Safety of Life at Sea
SSA	space situational awareness
SSR	Space Sustainability Rating

STC	space traffic coordination
STM	space traffic management
SWIFT	Society for Worldwide Interbank Financial Telecommunication
TCBM	transparency and confidence-building measure
UN	United Nations
UNCLOS	United Nations Conference on the Law of the Sea
UNCOPUOS	United Nations Committee on the Peaceful Uses of Outer Space
UNOOSA	United Nations Office for Outer Space Affairs
USSR	Union of Soviet Socialist Republics

References

AIAA—*See* American Institute of Aeronautics and Astronautics.

Ailor, William H., "Space Traffic Management: Implementations and Implications," *Acta Astronautica*, Vol. 58, No. 5, March 2006.

Antoni, Ntorina, Christina Giannopapa, and Kai-Uwe Schrogl, "Legal and Policy Perspectives on Civil–Military Cooperation for the Establishment of Space Traffic Management," *Space Policy*, Vol. 53, August 2020.

Article 19, "Internet: Content Moderation at Infrastructure Level Puts Rights at Risk," blog, October 25, 2021.

Barrow, Winford W., "Consideration of the New International Rules for Preventing Collisions at Sea," *Tulane Law Review*, Vol. 51, No. 4, 1976–1977, p. 1182.

Bennett, James, *The NORAD Experience: Implications for International Space Surveillance Data Sharing*, Secure World Foundation, August 1, 2010.

Bohan, Margaret, "NOAA's Participation in the U.S. Extended Continental Shelf Project" National Oceanic and Atmospheric Administration Office of Ocean Exploration and Research, undated.

Bouvé, Clement L., "The Development of International Rules of Conduct in Air Navigation," *Air Law Review*, Vol. 1, No. 1, January 1930.

Brittingham, Bryon C., "Does the World Really Need New Space Law?" *Oregon Review of International Law*, Vol. 12, No. 1, 2010.

Chatzipanagiotis, Michael, "Looking into the Future: The Case for an Integrated Aerospace Traffic Management," paper presented at the 58th IISL Colloquium on the Law of Outer Space, Jerusalem, Israel, October 2015.

Chow, Brian G., "How to Convince China and Russia to Join a Space Traffic Management Program for Peace and Prosperity," *SpaceNews*, January 26, 2021.

Chow, Brian G., "Space Traffic Management in the New Space Age," *Strategic Studies Quarterly*, Vol. 14, No. 4, Winter 2020.

Christensen, Ian A., and Christopher D. Johnson, "Putting the White House Executive Order on Space Resources in an International Context," *Space Review*, April 27, 2020

Ciccorossi, Jorge, "Harmful Interference to Satellite Systems and the Current Challenge to GNSS," paper presented at Eurocontrol Stakeholder Forum on GNSS, March 4, 2021.

Cioară, Octavian, *Aviation System Block Upgrade (ASBU) Implementation Monitoring Report*, International Civil Aviation Organization and Eurocontrol, 2021

Civil Aviation Authority, "ICAO's Structure and Upcoming Events in the Field of Civil Aviation (Infographics)," Government of Poland, June 11, 2013.

Clark, Julian, ed., *Shipping Laws and Regulations 2022–2023*, Global Legal Group, 2022.

Contant-Jorgenson, Corinne, Petr Lála, and Kai-Uwe Schrogl, eds., *Cosmic Study on Space Traffic Management*, International Academy of Astronautics, 2006.

Corbett, Julian, *The League of Nations and Freedom of the Seas*, Oxford University Press, London, 1918.

Cottom, Travis S., "Creating a Space Traffic Management System and Potential Geopolitical Opportunities," *Astropolitics*, Vol. 19, No. 1–2, May–August 2021.

Council of European Aerospace Societies, *1st CEAS European Air and Space Conference*, German Society for Aeronautics and Astronautics, 2007.

Cute, Brian, "Evolving ICANN's Multistakeholder Model: The Work Plan and Way Forward," ICANN Blogs, December 23, 2019.

De Moraes, Rodrigo Fracalossi, "The Parting of the Seas: Norms, Material Power and State Control over the Ocean," *Revista Brasileira de Política Internacional* [*Brasilian Review of International Politics*], Vol. 62, No. 1, April 15, 2019.

Dean, Paul, Tom Walters, Jonathon Goulding, and Henry Clack, *Autonomous Ships: MASS for the MASSes*, Holman Fenwick Willan, 2022.

Department of Defense Instruction 4540.1, *Use of International Airspace by U.S. Military Aircraft and for Missile and Projectile Firings*, change 1, May 22, 2017

Desautels, Eric, "U.S. Statement to the Open-Ended Working Group on Reducing Space Threats Through Norms, Rules and Principles of Responsible Behavior," U.S. Mission to International Organizations in Geneva, May 9, 2022.

Dominguez, Michael, Martin Faga, Jane Fountain, Patrick Kennedy, and Sean O'Keefe, *Space Traffic Management*, National Academy of Public Administration, August 2020.

Doucet, Gilles, "Characteristics of Governance for Domains Beyond National Jurisdiction: Lessons for Future Outer Space Governance," *Astropolitics*, Vol. 20, No. 2–3, May–December 2022.

Durand, Alain, *New IP*, Office of the Chief Technology Officer, International Corporation for Assigned Names and Numbers, October 27, 2020

Eichengreen, Barry, "Sanctions, SWIFT, and China's Cross-Border Interbank Payments System," Center for Strategic and International Studies, May 20, 2022.

Eilstrup-Sangiovanni, Mette, "Death of International Organizations. The Organizational Ecology of Intergovernmental Organizations, 1815–2015," *Review of International Organizations*, Vol. 15, No. 2, April 2020.

European Space Agency, "Space Debris by the Numbers," webpage, undated. As of April 6, 2023: https://www.esa.int/Space_Safety/Space_Debris/Space_debris_by_the_numbers

European Space Policy Institute, *Space Environment Capacity: Policy, Regulatory and Diplomatic Perspectives on Threshold-Based Models for Space Safety and Sustainability*, April 11, 2022.

European Union, "An EU Approach for Space Traffic Management," fact sheet, February 15, 2022.

EUSTM, "STM Is Getting a Prominent Place in European Space Policymaking," June 29, 2022.

Experimental Aircraft Association, "Transponder Requirements," webpage, undated. As of April 5, 2023: https://www.eaa.org/eaa/aircraft-building/intro-to-aircraft-building/frequently-asked-questions/transponder-requirements

Farrell, Maria, "Quietly, Symbolically, US Control of the Internet Was Just Ended," *The Guardian*, March 14, 2016.

Federal Aviation Administration, "Air Traffic Control," webpage, undated. As of April 5, 2023: https://www.faa.gov/about/history/photo_album/air_traffic_control

Federal Aviation Administration, "First Air Traffic Controller Remembered," webpage, undated. As of April 6, 2023:
https://www.faa.gov/news/communications/controller_remembered/

Federal Aviation Administration, *Introduction to TCAS II*, version 7.1, February 28, 2011.

Financial Stability Board, *Enhancing Cross-Border Payments: Stage 1 Report to the G20*, April 9, 2020.

Finnemore, Martha, and Duncan B. Hollis, "Constructing Norms for Global Cybersecurity," *American Journal of International Law*, Vol. 110, No. 3, July 2016.

Frandsen, Hjalte Osborn, "Looking for the Rules-of-the-Road of Outer Space: A Search for Basic Traffic Rules in Treaties, Guidelines and Standards," *Journal of Space Safety Engineering*, Vol. 9, No. 2, June 2022.

Fukuyama, Francis, "What Is Governance?" *Governance*, Vol. 26, No. 3, July 2013.

Fukuyama, Francis, "Governance: What Do We Know, and How Do We Know It?" *Annual Review of Political Science*, Vol. 18, December 21, 2015.

Garber, Stephen, and Marissa Herron, "How Has Traffic Been Managed in the Sky, on Waterways, and on the Road? Comparisons for Space Situational Awareness (Part 2)," *Space Review*, June 15, 2020.

Ghosh, Snehashish, and Anirudh Sridhar, "Internet Corporation for Assigned Names and Numbers (ICANN)," Centre for Internet and Society, undated.

Gibbs, Graham, and Ian Pryke, "International Cooperation in Space: The AIAA–IAC Workshops" *Space Policy*, Vol. 19, No. 1, February 2003.

Gillet, Emmanuel, "WhoIs Data: New Response from ICANN to the European Commission," *IP Twins*, April 25, 2022

Glass, Andrew, "Congress Passed Air Commerce Act, May 20, 1926," *Politico*, May 20, 2013.

Glavind, Johanne Grøndahl, "Bureaucratic Power at Play? The Performance of the EU in the International Atomic Energy Agency," *European Security*, Vol. 24, No. 1, 2015.

Gleason, Michael P., "Establishing Space Traffic Management Standards, Guidelines and Best Practices," *Journal of Space Safety Engineering*, Vol. 7, No. 3, September 2020.

Gleason, Michael P., and Travis Cottom, *U.S. Space Traffic Management: Best Practices, Guidelines, Standards, and International Considerations*, Aerospace Corporation, August 2018.

Global Fishing Watch, "What Is the Automatic Identification System (AIS)?" webpage, undated. As of April 6, 2023:
https://globalfishingwatch.org/faqs/what-is-ais/

Global Industry Analysts, *Global Space Situational Awareness (SSA) Industry*, January 2023.

Gray, Julia, "Life, Death, or Zombie? The Vitality of International Organizations," *International Studies Quarterly*, Vol. 62, No. 1, March 2018.

Griffin, Martin, "Integration of Aerospace Operations into the Global Air Traffic Management System," paper presented at the Space Traffic Management Conference 2014: Roadmap to the Stars, Daytona Beach, Fla., November 5–6, 2014.

Grotius, Hugo, *Mare Liberum [The Free Sea]*, trans. by Richard Hakluyt, Liberty Fund, [1609] 2004.

Grzelka, Zachary, and Jeffrey Wagner, "Managing Satellite Debris in Low-Earth Orbit: Incentivizing Ex Ante Satellite Quality and Ex Post Take-Back Programs," *Environmental and Resource Economics*, Vol. 74, 2019.

Gulrajani, Nilima, Sebastian Haug, and Silke Weinlich, "Fixing UN Financing: A Pandora's Box the World Health Organization Should Open," ODI, January 26, 2022

Guzman, Andrew, "The Consent Problem in International Law," working paper, University of California Berkley Program in Law and Economics, March 10, 2011.

HG Legal Resources, "What Is a Flag of Convenience?" webpage, undated. As of April 6, 2023: https://www.hg.org/legal-articles/what-is-a-flag-of-convenience-31395

Hitchens, Theresa, "Forwarding Multilateral Space Governance: Next Steps for the International Community," working paper, Center for International and Security Studies at Maryland, University of Maryland, August 2018.

Hitchens, Theresa, "At UN Meeting, Space Cooperation Picks up Momentum, but Moscow and Beijing Play Spoilers," *Breaking Defense*, February 3, 2023a.

Hitchens, Theresa, "Balloons vs. Satellites: Popping Some Misconceptions about Capability and Legality," *Breaking Defense*, February 7, 2023b.

Hoagland, P., J. Jacoby, and M. E. Schumacher, "Law of the Sea," in John H. Steele, ed., *Encyclopedia of Ocean Sciences*, 2nd ed., Elsevier, 2001.

Hunter, Stephen, "How to Reach an International Civil Aviation Organization Role in Space Traffic Management," paper presented at the Space Traffic Management Conference 2014: Roadmap to the Stars, Daytona Beach, Fla., November 5–6, 2014.

Hunter, Stephen, "Safe Operations Above FL600," paper presented at the Space Traffic Management Conference 2015: The Evolving Landscape, Daytona Beach, Fla., November 12–13, 2015.

ICANN—*See* International Corporation for Assigned Names and Numbers.

ICAO—*See* International Civil Aviation Organization.

ICAO/UNOOSA AeroSPACE Symposium, Montréal, Canada, March 18–20, 2015.

IMO—*See* International Maritime Organization.

Inmarsat, *Space Sustainability Report: Making the Case for ESG Regulation, International Standards and Safe Practices in Earth Orbit*, June 22, 2022.

International Academy of Astronautics, International Astronautical Federation, and International Institute of Space Law, "Cooperative Initiative to Develop Comprehensive Approaches and Proposals for Space Traffic Management (STM)," September 17, 2022.

International Civil Aviation Organization, "1928: The Havana Convention," webpage, undated-a. As of April 6, 2023: https://applications.icao.int/postalhistory/1928_the_havana_convention.htm

International Civil Aviation Organization, "About ICAO," webpage, undated-b. As of April 6, 2023: https://www.icao.int/about-icao/Pages/default.aspx

International Civil Aviation Organization, "Assembly," webpage, undated-c. As of April 6, 2023: https://www.icao.int/about-icao/assembly/Pages/default.aspx

International Civil Aviation Organization, "Freedoms of the Air," webpage, undated-d. As of April 6, 2023:
https://www.icao.int/pages/freedomsair.aspx

International Civil Aviation Organization, "The History of ICAO and the Chicago Convention," webpage, undated-e. As of April 6, 2023:
https://www.icao.int/about-icao/History/Pages/default.aspx

International Civil Aviation Organization, "The ICAO Council," webpage, undated-f. As of April 6, 2023:
https://www.icao.int/about-icao/Council/Pages/council.aspx

International Civil Aviation Organization, "ICAO Secretariat," webpage, undated-g. As of April 6, 2023:
https://www.icao.int/secretariat/Pages/default.aspx

International Civil Aviation Organization, "Introduction," webpage, undated-h. As of April 6, 2023:
https://www.icao.int/ChicagoConference/Pages/chicago-conference-introduction.aspx

International Civil Aviation Organization, "Milestones in International Civil Aviation," webpage, undated-i. As of April 6, 2023:
https://www.icao.int/about-icao/History/Pages/Milestones-in-International-Civil-Aviation.aspx

International Civil Aviation Organization, "Organizations Able to Be Invited to ICAO Meetings," webpage, undated-j. As of April 6, 2023:
https://www.icao.int/about-icao/Pages/Invited-Organizations.aspx

International Civil Aviation Organization, "Communications, Navigation, and Surveillance (CNS) Section," webpage, undated-k. As of April 6, 2023:
https://www.icao.int/safety/pages/cns.aspx

International Civil Aviation Organization, "Overview of Automatic Dependent Surveillance-Broadcast (ADS-B) Out," briefing slides, undated-l.

International Civil Aviation Organization, "The Paris Convention of 1910: The Path to Internationalism," webpage, undated-l. As of April 6, 2023:
https://applications.icao.int/postalhistory/1910_the_paris_convention.htm

International Civil Aviation Organization, *Convention on International Civil Aviation*, Document 7300, December 7, 1944.

International Civil Aviation Organization, *Synopsis*, Uberlingen Accident Report, 2002.

International Civil Aviation Organization, *ADSB Implementation and Operations Guidance Document*, 7th ed., September 2014.

International Civil Aviation Organization, "Making an ICAO SARP," factsheet, March 5, 2018.

International Civil Aviation Organization, "Budget of the Organization for 2020-2021-2022," webpage, 2019. As of April 13, 2023:
https://www.icao.int/annual-report-2019/Pages/
supporting-strategies-finances-budget-2020-2021-2022.aspx

International Civil Aviation Organization, "Progress on ICAO Accountability and Transparency," webpage, 2021. As of April 13, 2023:
https://www.icao.int/annual-report-2021/Pages/
supporting-strategies-human-resources-management-and-gender-equity.aspx

International Civil Aviation Organization, *Manual on the Development of Regional and National Aviation Safety Plans*, Document 10131, 2nd ed., 2022.

International Civil Aviation Organization [@ICAO], "ICAO is strongly concerned by the apparent forced landing of a Ryanair Flight and its passengers, which could be in contravention of the Chicago Convention. We look forward to more information being officially confirmed by the countries and operators concerned," Twitter post, May 23, 2021.

International Institute of Space Law, "IISL, IAA and IAF Conclude Major Report on STM— International Institute of Space Law," undated.

International Maritime Organization, "AIS Transponders," webpage, undated-a. As of April 6, 2023:
https://www.imo.org/en/OurWork/Safety/Pages/AIS.aspx

International Maritime Organization, "Autonomous Shipping," webpage, undated-b. As of April 6, 2023:
https://www.imo.org/en/MediaCentre/HotTopics/Pages/Autonomous-shipping.aspx

International Maritime Organization, "Convention on the International Maritime Organization," webpage, undated-c. As of April 6, 2023:
https://www.imo.org/en/About/Conventions/Pages/Convention-on-the-International-Maritime-Organization.aspx

International Maritime Organization "Convention on the International Regulations for Preventing Collisions at Sea, 1972 (COLREGs)," webpage, undated-d. As of April 6, 2023:
https://www.imo.org/en/About/Conventions/Pages/COLREG.aspx

International Maritime Organization, "Frequently Asked Questions," webpage, undated-e. As of April 6, 2023:
https://www.imo.org/en/About/Pages/FAQs.aspx

International Maritime Organization, "International Convention for the Safety of Life at Sea (SOLAS), 1974," webpage, undated-f. As of April 6, 2023:
https://www.imo.org/en/About/Conventions/Pages/International-Convention-for-the-Safety-of-Life-at-Sea-(SOLAS),-1974.aspx

International Maritime Organization, "Introduction to IMO," webpage, undated-g. As of April 6, 2023:
https://www.imo.org/en/About/Pages/Default.aspx

International Maritime Organization, "Safety of Navigation," webpage, undated-h. As of April 6, 2023:
https://www.imo.org/en/OurWork/Safety/Pages/NavigationDefault.aspx

International Maritime Organization, "Structure of IMO," webpage, undated-i. As of April 6, 2023:
https://www.imo.org/en/About/Pages/Structure.aspx

International Maritime Organization, *Status of IMO Treaties*, September 29, 2021.

International Maritime Organization, *Financial Report and Audited Financial Statements for the Year Ended 31 December 2021*, December 31, 2021.

International Organization for Standardization, "About ISO 20022," webpage, undated. As of April 6, 2023:
https://www.iso20022.org/about-iso-20022

International Relations and Defence Committee, *UNCLOS: The Law of the Sea in the 21st Century*, House of Lords, March 1, 2022.

International Seabed Authority, "Equitable Sharing of Financial and Other Economic Benefits from Deep-Sea Mining," January 2022.

International Space University, *Space Traffic Management*, 2007.

International Telecommunication Union, "History," webpage, undated-a. As of April 28, 2023:
https://www.itu.int/en/about/Pages/history.aspx

International Telecommunication Union, "What Does ITU Do?" webpage, undated-b. As of April 6, 2023:
https://www.itu.int:443/en/about/Pages/whatwedo.aspx

International Telecommunication Union, "How Is ITU Funded?" webpage, updated May 2022. As of April 6, 2023:
https://www.itu.int:443/en/mediacentre/backgrounders/Pages/how-is-itu-funded.aspx

International Telecommunication Union, "ITU-T Standardization: Committed to Connecting the World," 2010.

International Tribunal for the Law of the Sea, "Latest News," webpage, undated. As of April 6, 2023:
https://www.itlos.org/en/main/latest-news/

Internet Corporation for Assigned Names and Numbers, "ICANN Expected Standards of Behavior," webpage, undated-a. As of April 6, 2023:
https://www.icann.org/resources/pages/expected-standards-2016-06-28-en

Internet Corporation for Assigned Names and Numbers, "ICANN'S Early Days," webpage, undated-b. As of April 6, 2023:
https://www.icann.org/en/history/early-days

Internet Corporation for Assigned Names and Numbers, "What Does ICANN Do?" webpage, undated-c. As of April 6, 2023:
https://www.icann.org/resources/pages/what-2012-02-25-en

Internet Corporation for Assigned Names and Numbers, "Bylaws for Internet Corporation for Assigned Names and Numbers: A California Nonprofit Public-Benefit Corporation–ICANN," webpage, updated June 2, 2022. As of April 6, 2023:
https://www.icann.org/resources/pages/governance/bylaws-en

Jacobsson, Marie, "Institutional Arrangements for the Ocean: From Zero to Indefinite?" *Ecology Law Quarterly*, Vol. 46, No. 1, March 31, 2019.

Jakhu, Ram S., Tommaso Sgobba, and Paul Stephen Dempsey, eds. *The Need for an Integrated Regulatory Regime for Aviation and Space: ICAO for Space?* Springer, 2011.

Kaul, Sanat, "Integrating Air and Near Space Traffic Management for Aviation and Near Space," *Journal of Space Safety Engineering*, Vol. 6, No. 2, June 2019.

Koga, Kei, "A New Strategic Minilateralism in the Indo-Pacific," *Asia Policy*, Vol. 17, No. 4, October 2022

Lal, Bhavya, Asha Balakrishnan, Becaja M. Caldwell, Reina S. Buenconsejo, and Sara A. Carioscia, *Global Trends in Space Situational Awareness (SSA) and Space Traffic Management (STM)*, Institute for Defense Analysis, 2018.

Larsen, Paul, "Space Traffic Management Standards," *Journal of Air Law and Commerce*, Vol. 83, No. 2, 2018.

Leal-Arcas, Rafael, "Top-Down and Bottom-Up Approaches in Climate Change and International Trade," paper presented at the 10th Annual Conference of the Euro-Latin Study Network on Integration and Trade (ELSINT), October 19–20, 2012.

Lebryk, Theo, "The Fight over the Fate of the Internet: The Economic, Political, and Security Costs of China's Digital Standards Strategy," *China Focus*, April 21, 2021.

Legal Subcommittee 2015, United Nations Office for Outer Space Affairs, April 13–25, 2015.

Lemoine, Derek, "Incentivizing Negative Emissions Through Carbon Shares," working paper, National Bureau of Economic Research, November 2021.

Ligor, Douglas C., Benjamin Miller, Maria McCollester, Brian Phillips, Geoffrey Kirkwood, Josh Becker, Gwen Mazzotta, Bruce McClintock, and Barbara Bicksler, *Assessing the Readiness for Human Commercial Spaceflight Regulations: Charting a Trajectory from Revolutionary to Routine Travel*, RAND Corporation, RR-A2466-1, 2023. As of April 6, 2023: https://www.rand.org/pubs/research_reports/RRA2466-1.html

Looper, Matthew G., "International Space Law: How Russia and the U.S. Are at Odds in the Final Frontier," *South Carolina Journal of International Law and Business*, Vol. 18, No. 2, 2022.

MacDonald, Bruce W., Carla P. Freeman, and Alison McFarland, *China and Strategic Instability in Space: Pathways to Peace in an Era of US-China Strategic Competition*, United States Institute of Peace, February 9, 2023.

MacKenzie, David, *ICAO: A History of the International Civil Aviation Organization*, University of Toronto Press, 2010.

Mahan, Alfred Thayer, *The Influence of Sea Power Upon History 1660–1783,* 12th ed., Project Gutenberg, [1890] 2004.

McClintock, Bruce, Katie Feistel, Douglas C. Ligor, and Kathryn O'Connor, *Responsible Space Behavior for the New Space Era: Preserving the Province of Humanity*, RAND Corporation, PE-A887-2, 2021. As of April 6, 2023: https://www.rand.org/pubs/perspectives/PEA887-2.html

McCormick, Dan, Douglas C. Ligor, and Bruce McClintock, *Cross-Domain Lessons for Space Traffic: An Analysis of Air and Maritime Treaty Governance Mechanisms*, RAND Corporation, RR-A2208-2, 2023. As of April 6, 2023: https://www.rand.org/pubs/research_reports/RRA2208-2.html

McDowell, Jonathan, *Space Activities in 2022*, Jonathan's Space Report, January 17, 2023.

Mendelsohn, Allan I., "Flags of Convenience: Maritime and Aviation," *Journal of Air Law and Commerce*, Vol. 79, No. 1, Winter 2014.

Miller, Steven, *Hard Times for Arms Control: What Can Be Done?* The Hague Center for Strategic Studies, February 2022.

Montgomery, Mark, and Theo Lebryk, "China's Dystopian 'New IP' Plan Shows Need for Renewed US Commitment to Internet Governance," *Just Security*, April 13, 2021.

Muelhaupt, Theodore J., Marlon E. Sorge, Jamie Morin, and Robert S. Wilson, "Space Traffic Management in the New Space Era," *Journal of Space Safety Engineering*, Vol. 6, No. 2, June 2019.

National Geographic, "Jun 7, 1494 CE: Treaty of Tordesillas," updated October 4, 2022.

NATO Shipping Centre, "AIS (Automatic Identification System) Overview," NATO, 2021.

Norris, Andrew J., "The 'Other' Law of the Sea," *Naval War College Review*, Vol. 64, No. 3, Summer 2011.

O'Connell, Daniel Patrick, "The History of the Law of the Sea," in Ivan Anthony Shearer, ed., *The International Law of the Sea*, Vol. 1, 1st ed., Clarendon Press, 1982.

Organisation for Economic Co-operation and Development, *What Have We Learned About Extended Producer Responsibility in the Past Decade? Case Study—Chile*, 2014.

Oltrogge, Daniel L., "The 'We' Approach to Space Traffic Management," paper presented at the 2018 SpaceOps Conference, American Institute of Aeronautics and Astronautics, May 28–June 1, 2018.

Paris Declaration Respecting Maritime Law, April 16, 1856.

Perek, Lubos, "Traffic Rules for Outer Space," *Proceedings of the Twenty-Fifth Colloquium on the Law of Outer Space*, American Institute of Aeronautics and Astronautics, September–October 1982.

Peterson, Glenn, Marlon Sorge, and William Ailor, *Space Traffic Management in the Age of New Space*, Aerospace Corporation, April 2018.

Pevehouse, Jon C. W., Timothy Nordstrom, Roseanne W. McManus, and Anne Spencer Jamison, "Tracking Organizations in the World: The Correlates of War IGO Version 3.0 Datasets," *Journal of Peace Research*, Vol. 57, No. 3, June 2020.

Pultarova, Tereza, "SpaceX Starlink Satellites Responsible for over Half of Close Encounters in Orbit, Scientist Says," *Space.com*, August 20, 2021.

Rainbow, Jason, "Connecting the Dots: Improving Satellite Collision Predictions for Efficient Space," *SpaceNews*, June 23, 2022.

Rao, Akhil, Matthew G. Burgess, and Daniel Kaffine, "Orbital-Use Fees Could More than Quadruple the Value of the Space Industry," *Proceedings of the National Academy of Sciences*, Vol. 117, No. 23, June 9, 2020.

Rathgeber, Wolfgang, Kai-Uwe Schrogl, and Ray Williamson, eds., *The Fair and Responsible Use of Space: An International Perspective*, Springer, 2010.

Rouillon, Sébastien, "A Physico-Economic Model of Low Earth Orbit Management," *Environmental and Resource Economics*, Vol. 77, No. 4, December 2020.

Sadat, Mir, and Julie Siegel, *Space Traffic Management: Time for Action*, Atlantic Council, August 2022.

S., L., "Why Is America Giving up Control of ICANN?" *The Economist*, September 30, 2016.

Scholte, Jan Aart, Soetkin Verhaegen, and Jonas Tallberg, "Elite Attitudes and the Future of Global Governance." *International Affairs*, Vol. 97, No. 3, May 2021.

Schrogl, Kai-Uwe, Corinne Jorgenson, Jana Robinson, and Alexander Soucek, eds., *Space Traffic Management—Towards a Roadmap for Implementation*, International Academy of Astronautics, 2018.

Scott, Susan, and Markos Zachariadis, "Origins and Development of SWIFT, 1973–2009," *Business History*, Vol. 54, No. 3, June 2012.

Scott, Susan, and Markos Zachariadis, *The Society for Worldwide Interbank Financial Telecommunication (SWIFT): Cooperative Governance for Network Innovation, Standards, and Community*, Routledge, 2014.

Seyer, Sean, *Sovereign Skies: The Origins of American Civil Aviation Policy*, Johns Hopkins University Press, 2021.

Smith II, George P., "The Politics of Lawmaking: Problems in International Maritime Regulation: Innocent Passage v. Free Transit," *University of Pittsburgh Law Review*, Vol. 37, No. 3, 1976.

Society for Worldwide Interbank Financial Telecommunication, *ISO 20022 for Dummies*, Wiley, 2020.

Society for Worldwide Interbank Financial Telecommunication, "Swift Governance," webpage, undated-a. As of April 6, 2023:
https://www.swift.com/about-us/organisation-governance/swift-governance

Society for Worldwide Interbank Financial Telecommunication, "Swift Oversight," webpage, undated-b. As of April 6, 2023:
https://www.swift.com/about-us/organisation-governance/swift-oversight

Space Law Symposium 2015, International Institute of Space Law and European Centre for Space Law, April 13, 2015.

Space Traffic Management Conference 2019: Progress Through Collaboration, Embry-Riddle Aeronautical University, February 26–27, 2019.

Stilwell, Ruth E., "Decentralized Space Traffic Management," paper presented at Space Traffic Management Conference 2019: Progress Through Collaboration, Daytona Beach, Fla., February 26–27, 2019.

Stilwell, Ruth E., "Who Is Right When It Comes to the Right of Way in Space?" paper presented at Facing the Security Challenge, 6th Annual Space Traffic Management Conference, University of Texas, February 19–20, 2020.

Stilwell, Ruth E., Diane Howard, and Sven Kaltenhauser, "Overcoming Sovereignty for Space Traffic Management," *Journal of Space Safety Engineering*, Vol. 7, No. 2, June 2020.

Stockton, Charles H., "The Declaration of Paris," *American Journal of International Law*, Vol. 14, No. 3, July 1920.

Takeuchi, Yu, "STM in the Nature of International Space Law," paper presented at Space Traffic Management Conference 2019: Progress Through Collaboration, Daytona Beach, Fla., February 26–27, 2019.

Tan, Huileng, "China and Russia Are Working on Homegrown Alternatives to the SWIFT Payment System. Here's What They Would Mean for the US Dollar," *Insider*, April 28, 2022.

Treves, Tullio, "1958 Geneva Conventions on the Law of the Sea," Audiovisual Library of International Law, September 2008.

Treves, Tullio, "Historical Development of the Law of the Sea" in Donald Rothwell, Alex Oude Elferink, Karen Scott, and Tim Stephens, eds., *Oxford Handbook of the Law of the Sea*, Oxford University Press, 2015.

Tublin, Melvin, "The New Danger Signal Authorized by the International Rules of the Road," *Georgetown Law Journal*, Vol. 42, No. 1, November 1953.

UN—*See* United Nations.

United Arab Emirates Ministry of Energy & Infrastructure, "International Maritime Organization," webpage, undated. As of May 16, 2023:
https://www.moei.gov.ae/en/about-ministry/imo.aspx

United Nations, "5 Things You Should Know About ICAO, the UN Aviation Agency," UN News, May 26, 2021.

United Nations, "Meetings of States Parties to the 1982 Convention on the Law of the Sea," webpage, March 15, 2022. As of April 6, 2023:
https://www.un.org/depts/los/meeting_states_parties/meeting_states_parties.htm

United Nations, "Open-Ended Working Group on Reducing Space Threats," webpage, 2022. As of April 13, 2023:
https://meetings.unoda.org/open-ended-working-group-on-reducing-space-threats-2022

United Nations, "Reducing Space Threats Through Norms, Rules, and Principles of Responsible Behaviours," resolution, A/RES/75/36, December 7, 2020.

United Nations Convention on the Law of the Sea, December 10, 1982.

United Nations Committee on the Peaceful Uses of Outer Space, Legal Subcommittee , "2016 LSC Draft Report," United Nations Office for Outer Space Affairs, April 7, 2016.

United Nations Office for Outer Space Affairs, "Proposal for a Single Issue/Item for Discussion at the Fifty-Fifth Session of the Legal Subcommittee in 2016 on: 'Exchange of Views on the Concept of Space Traffic Management,'" April 14, 2015.

U.S. Coast Guard, "USCG IMO Homepage." webpage, updated February 6, 2018. As of April 6, 2023:
https://www.dco.uscg.mil/IMO/International-Maritime-Organization-Sub-committees-HWT/

Van Roste, Peter, "ICANN71: CcTLD Governance Models—Why One Size Does Not Fit All," Council of European National Top-Level Domain Registries, blog, June 16, 2021.

Von Borzyskowski, Inken, and Felicity Vabulas, "Hello, Goodbye: When Do States Withdraw from International Organizations?" *Review of International Organizations*, Vol 14, No. 2, June 2019.

Wahal, Anya, "On International Treaties, the United States Refuses to Play Ball," *The Internationalist*, blog, Council on Foreign Relations, January 7, 2022.

Weeden, Brian, "Muddling Through Space Traffic Management," *SpaceNews*, September 22, 2017.

Weitzenboeck, Emily M., "Hybrid Net: The Regulatory Framework of ICANN and the DNS," *International Journal of Law and Information Technology*, Vol. 22, No. 1, Spring 2014.

Wong, Liana, and Rebecca M. Nelson, *International Financial Messaging Systems*, Congressional Research Service, R46843, July 19, 2021.

Yu Hongyuan and Yu Bowen, "Global Climate Governance: New Trends and China's Policy Options," *China International Studies*, Vol. 61, December 12, 2016.

Zhao Yun, "Initial Thoughts on a Possible Regime for Space Traffic Management," *Centre for Aviation and Space Laws*, blog, June 14, 2022.